Science Networks · Historical Studies
Volume 5

Birkhäuser Verlag
Basel · Boston · Berlin

Nahum Kipnis

History of the Principle of Interference of Light

1991

Birkhäuser Verlag
Basel · Boston · Berlin

Nahum Kipnis
The Bakken
3537 Zenith Ave. So.
Minneapolis, MN 55416
USA

Library of Congress Cataloging in Publication Data

Kipnis, Nahum S.
History of the principle of interference of light/Nahum Kipnis.
p. cm. – (Science networks historical studies; v. 5)
Revision of thesis (Ph. D.) – University of Minnesota, 1984.
Includes bibliographical references.
ISBN 0-8176-2316-7 (U.S.)
1. Interference (Light) – History. 2. Optics – History – 19th century. 3. Young, Thomas, 1773–1829. I. Title. II. Series.
QC411.K53 1990
535.4 – dc20

CIP-Titelaufnahme der Deutschen Bibliothek

Kipnis, Nahum:
History of the principle of interference of light/Nahum Kipnis. – Basel; Boston; Berlin: Birkhäuser, 1990
(Science networks historical studies; Vol. 5)
ISBN 3-7643-2316-7 (Basel…)
ISBN 0-8176-2316-7 (Boston)
NE: GT

Printed in Germany on acid-free paper
ISBN 3-7643-2316-7
ISBN 0-8176-2316-7

Contents

To Berta

Preface

The controversy between the wave theory and the emission theory of light early in the nineteenth century has been a subject of numerous studies. Yet many issues remain unclear, in particular, the reasons for rejecting Young's theory of light. It appears that further progress in the field requires a better grasp of the overall situation in optics and related subjects at the time and a more thorough study of every factor suggested to be of importance for the dispute.

This book is intended to be a step in this direction. It examines the impact of the concept of interference of light on the development of the early nineteenth-century optics in general, and the theory of light, in particular. This is not a history of the wave theory of light, nor is it a history of the debate on the nature of light in general: it covers only that part of the controversy which involved the concept of interference. Although the book deals with a number of scientists, scientific institutions, and journals, its main character is a scientific concept, the principle of interference. While discussing the reasons for accepting or rejecting this concept I have primarily focused on scientific factors, although in some cases the human factor is examined as well.

The book is a revised Ph.D. dissertation (University of Minnesota, 1984) written under Alan E. Shapiro. My first gratitude is to the late Usher Frankfurt (Moscow) who introduced me to the history of science. I am indebted to Alan Shapiro and Roger Stuewer for helping me to settle in the U.S.A. and continue my research as well for providing examples of fine scholarship and teaching. Ivor Grattan-Guinness and Charles Gillispie deserve my special appreciation for their efforts to get my thesis published.

I would like to thank the librarians and staff of the following institutions: Interlibrary Loan Division of the University of Minnesota Libraries, British Library, the Royal Society Library, the Library of University College London, and Bibliothèque de l'Institut.

For permitting me to reproduce manuscript material I express my gratitude to the Royal Society, the Library of University College London, British Library, and Bibliothèque de l'Institut. I am grateful to the Minneapolis Foundation for a grant-in-aid of research.

My thanks go to John Senior, director, and my colleagues at The Bakken, a Library and Museum of Electricity in Life, for allowing me time and providing various help during my work on the book.

It is the most pleasant duty to thank my wife Berta for her patience and constant support and also for her help with drawings.

March 1989 Minneapolis N. Kipnis

Chapter I

The Concept of interference: Historiography and physics

I.1 The "mystery" of Young's theory

The debate on the nature of light began in antiquity and intensified in the second half of the seventeenth century. The major struggle was between the wave hypothesis and the emission (or corpuscular) hypothesis. Among the prominent scientists, Robert Hooke (1635–1703) and Christiaan Huygens (1629–1695) advocated the wave hypothesis, while Isaac Newton (1642–1727) was a champion of the emission hypothesis. Throughout the eighteenth century, the "emissionists" had the upper hand, and despite Leonhard Euler's (1707–1783) efforts to reverse this trend, by 1800, the wave hypothesis was almost universally ignored.

In 1799, Thomas Young (1773–1829) began to bring in new arguments in favor of the wave hypothesis of light. The most important one was the concept of interference of light, which he discovered in 1801. The new wave theory received little recognition, however, and the situation began to change only after Augustin-Jean Fresnel (1788–1827) rediscovered in 1815 the concept of interference and offered a theory very similar to Young's.

The fact that Fresnel's theory won acceptance while Young's theory did not has puzzled historians of science ever since the 1830s. They have concurred that the extreme conciseness of Young's papers could have prevented readers from understanding his theory but have disagreed on the role of other factors. Young's contemporaries François Arago (1786–1853), William Whewell (1794–1866), and George Peacock (1791–1858) argued that Young's theory was superior to the emission theory of light and was rejected for external reasons, such as Newton's opposition to the wave hypothesis of light and a severe criticism of Young's works by Henry Brougham (1778–1868).[1] This "externalist" view has been

[1] Arago, "Thomas Young", *Oeuvres completes*, 15 vols (Paris, 1954–59), 1: 241–94; Whewell, *History of the Inductive Sciences from the Earliest to the Present Time* [1837], 3rd ed., 2 vols (New York, 1901), 2: 111–18; Peacock, *Life of Thomas Young* (London, 1855), 174–85. In this book, either "wave theory" or "emission theory" means a quantitative theory based on the wave or emission hypothesis, respectively.

adopted by many modern scholars,[2] with a few new factors added, such as the strength of the emission theory,[3] and the lack of interest in optics in the early 1800s.[4]

Recently, several historians began to emphasize the crucial role of internal causes. According to John Worrall, the wave program degenerated in Young's hands.[5] Edgar Morse and Kenneth Latchford believe that Young's theory had no chance of winning because of an abundance of hypotheses, a dearth of observations, and primitive mathematics.[6] Frank James explains its failure by lack of a proper physical model of the interaction of light and matter.[7]

There is no agreement, however, about the role of such internal factor as the concept of interference. Geoffrey Cantor suggested that this concept could have helped Young had his contemporaries not "overlooked" it[8] and thus rejected his theory, "as a mere restatement of Euler's theory."[9] According to Latchford and Worrall, however, Young's concept of interference was so poorly presented that it could have hindered the acceptance of his wave theory rather than accelerated it.[10]

Since little evidence, if any, has been given to support all these points, they remain suggestions waiting for a thorough investigation. Moreover, it seems that there could have been other causes affecting the debate on the nature of light in the early nineteenth century, for instance, the influence of related sciences, practical needs, changes in philosophy of science, etc. A complete solution of the problem would require a very extensive investigation, and my objective in this book is limited to studying the role of the principle of interference of light. Other factors will be examined only to the extent of their connection with the principle of interference.

---

2 E. Mach, *The Principles of Physical Optics* (London, 1926), 275–6; E. Whittaker, *A History of Theories of Aether and Electricity*, vol. 1 (New York, 1913), 102; A. Wood, *Thomas Young Natural Philosopher, 1773–1829* (Cambridge, 1954), 168–75.

3 H. Steffens, *The Development of Newtonian Optics in England* (New York: Science History Publications, 1977), 136.

4 G. Cantor, *Optics After Newton. Theories of Light in Britain and Ireland, 1704–1840* (Manchester University Press, 1983), 142.

5 Worrall, "Thomas Young and the 'refutation' of Newtonian optics," in C. Howson, ed., *Method and Appraisal in the Physical Sciences* (Cambridge, 1976), 143–9, 173. Hereafter cited as Worrall.

6 Latchford, "Thomas Young and the evolution of the interference principle," (Ph.D. diss., University of London, 1975), 3, 181, hereafter is referred to as Latchford; Morse, "Natural philosophy, hypotheses, and impiety: Sir David Brewster confronts the undulatory theory of light," (Ph.D diss., University of California, Berkeley, 1972), 111, 124.

7 James, "The physical interpretation of the wave theory of light," *British Journal for the History of Science* 17 (1984), 50–53.

8 Cantor, "Was Thomas Young a wave theorist?," *American Journal of Physics* 52 (1984), 308.

9 Cantor, *Optics After Newton*, 200.

10 Latchford, 33, 96, 101; Worrall, 137–140, 150–55.

To establish how the presence of a new component in the wave theory of light, the principle of interference, affected the reaction to the theory, it is necessary to separate the reception of the principle itself. One of the ways to do this is by checking how the principle of interference was met in acoustics where it was introduced by Young in 1800. Since the wave nature of sound was undisputed it could not adversely affect the perception of the new principle. Latchford already discussed in detail the origin of the acoustical principle of interference and found a connection between the responses in acoustics and optics. However, I found some of his conclusions contradicting historical facts and reexamined the question. It appears that a number of differences between Young scholars about the reception of his theory originate from deficiencies in methodology.

First, they tend to focus on a single cause, or on a very few causes of one type, "external" or "internal". Secondly, when trying to determine whether the wave theory was superior to the emission theory they rely on modern criteria of a scientific theory. Apparently, the assumption is that the contemporary criteria, which are unknown, have not changed; however such a view has no foundation whatsoever. Thirdly, historians believe that an assemblage of comments on Young's theory, taken at their face value and given the same weight, may provide (with rare exceptions) a true notion of the theory. This ignores the possibility that the critic's own views as well as personality could influence his judgement and produce a distorted picture of the theory itself and of its status. Finally, there is a great confusion about who accepted the wave theory and when, because the meaning of the terms "acceptance of a theory" and the "wave theory" is not stated.

To overcome these difficulties I developed a new methodology for studying the reception of the principle of interference. It may be also applicable to other theories, which did not enjoy an open discussion in scientific monographs and articles or an opposition of alternative theories which it brought to life; in short, when the response is mostly hidden and must be uncovered.

I.2 New approach

The new approach is based on the following assumptions. First, the acceptance of a theory depends on *many* factors, both external and internal. Secondly, the reaction to a given theory can be understood by comparing it with responses to other contemporary theories and similar theories offered subsequently. Thirdly, the role of internal factors can be determined by comparing the given theory to those other theories as to their structure, hypotheses, mathematics, and experimental support. Fourthly, to evaluate the role of such an external factor as critic's views and his personality one must study them separately, for instance, by comparing the critic's comment on the given subject with his comments on other subjects. Different opinions should be given different weight, according to their possible effect on that aspect of the response which is the target of the historical investigation.

The book is planned according to the idea of comparing the discovery of the concept of interference, its development, application, and reception in the cases of Young and Fresnel: the first six chapters deal with Young and his predecessors, and the last two are devoted to Fresnel. I covered only that part of Fresnel's theory which is relevant to the goal of this book, and the reader may consult Silliman and Buchwald about its other parts and a general overview.[11] Finally, to avoid confusion in judging the acceptance of a theory one should specify the content of the theory and the meaning of the term "acceptance of a theory."

Comparing Young's and Fresnel's concepts of interference raises two major difficulties. One is that many comments on their theories did not single out this concept. Still, such comments can be utilized if we can be certain that other parts of the wave theory did not overshadow the one based on the concept of interference, which will be called *the theory of interference*. This part, identical in Young's and Fresnel's early theories, was created to explain the phenomena of the so-called "periodical colors" (in modern terms, alternating colored fringes due to interference of reflected, refracted, and diffracted light). After 1817, however, Fresnel introduced new concepts such as the Huygens-Fresnel principle and the transversality of light waves, which led to a modification of his theory of interference and construction of new theories (for instance, the theory of double refraction). It may be shown, however (see Ch. VIII), that up to 1827, the impact of these new ideas was negligible.

Another is that in some of their judgements on the wave theory and the principle of interference historians contradict one another simply because they put different meanings into such terms as "the wave theory," "the concept of interference," the "acceptance of a theory," and others. To avoid confusion, it is necessary to specify these terms.

[11] Silliman, "Augustin Fresnel (1788–1827) and the establishment of the wave theory of light" (Ph.D. diss., Princeton University, 1968); Jed Buchwald, *The Rise of the Wave Theory of Light* (Chicago: University of Chicago Press, 1989), Chs. 5–8.

I.3 What is "Young's wave theory"?

It has been presumed that the progress in physical optics from the seventeenth century through the early nineteenth century depended primarily on the development of the debate on the nature of light. (By "progress" I mean the discovery of new phenomena and new laws.) According to some authors, it was also affected by a broader controversy between Cartesians and Newtonians.[12] The aspect frequently neglected, however, is that whether they embraced particles and forces or the aether and waves, the eighteenth-century scientists usually limited their debate on the nature and propagation of light to qualitative arguments and qualitative experiments. To some extent, Young inherited the tradition of building qualitative systems, which would have explained all physical phenomena; this, as Cantor demonstrated, is especially clear from Young's usage of the aether.[13] I intend to concentrate, however, on another aspect of his work, which was not typical for his time and which has not attracted attention of historians: a tendency to quantify physical optics. Thus, I want to explore how Young and his contemporaries viewed his theory of light in the framework of the controversy on *qualitative* versus *quantitative* approach to studying nature. In particular, I intend to prove that Young considered the concept of interference to be a basis for mathematizing his wave theory: "no man had ever attempted to advance a theory which would bear to be compared *mathematically* with the phenomena that I enumerated."[14] Henceforth, when speaking of "mathematizing" physical optics in general or a specific optical theory I mean *any* mathematical means. Thus, I will call "Young's wave theory" only its mathematical part. This part deals with periodical colors, and for reasons explained further it will be named the *theory of interference.*

I.4 "Interference" or "superposition"?

Since Young and his predecessors used different words to describe the effects of intersecting waves, we can compare their contributions by using modern terms,

[12] P. A. Pav, "Eighteenth century optics: The age of unenlightenment," (Ph.D. diss., Indiana University, 1964).

[13] Cantor, "The changing role of Young's ether," *British Journal for the History of Science* 5 (1970), 44–62.

[14] Young, *Reply to the Animadversions of the Edinburgh Reviewers on Some Papers in the Philosophical Transactions* (London, 1804). Here the references are given from *Miscellaneous Works of the late Thomas Young* . . . 3 vols (London, 1855), 1: 204. Hereafter is referred to as "Reply"

such as "interference", introduced by Young, and "superposition", of a later origin. Unfortunately, there is no agreement among historians on how to interpret these terms. For instance, to Worrall, both terms apparently mean the same thing, for he is constantly speaking of "superposition *or* interference."[15] Latchford reserves "interference" for Young's optics and "superposition" for his acoustics.[16] In his view, Young's term "interference" was an unfortunate one, since intersecting waves do not interfere with one another.[17] This opinion has been upheld by some physicists too.[18] In general, physicists are very vague about the meaning of "interference."[19] For this reason, I had to create myself a set of definitions, which reflect a physicist's view but are also adapted for an historical investigation.

First of all, I will introduce *several* "principles of superposition." The most general of them will be *the principle of superposition of motions* (in old terms, "combination of motions"): when a particle participates simultaneously in two motions, its compound instantaneous displacement equals a vector sum of the displacements, which the particle would have had if each motion existed alone. The first part of this formulation is the composition rule, while the second one defines independent motions. In other words, the principle of superposition is applicable only to independent motions.

Not all motions are independent. It has been known since antiquity that uniform motions are such. Daniel Bernoulli (1700–1782) found that vibrations of small amplitude are also independent. Thus, we may formulate a separate *principle of superposition of vibrations:* when two small vibrations coexist in the same point, the displacement of its compound motion can be found by applying the principle of superposition of motions to *every* moment of time, which means adding (analytically or graphically) the functions representing the component vibrations. A superposition of vibrations produces the best observable physical effect when the frequencies of two vibrations are either equal or very close. Let x_1 and x_2 be the displacements of two vibrations, a_1 and a_2 their amplitudes, A the amplitude of compound vibrations, ω_1 and ω_2 their frequencies, and α_1 and α_2

---

[15] Worrall, 137–9.

[16] Latchford, 32, 132, 151.

[17] Ibid., 33, 73.

[18] F. Jenkins & H. White, *Fundamentals of Optics*, 3rd ed. (New York, 1957), 232; R. S. Longhurst, *Geometrical and Physical Optics*, 2nd ed. (London: Longmans, 1967), 97.

[19] W. E. Williams, *Applications of Interferometry* (London, 1930), 1; R. G. Fowler & D. Meyer, *Physics for Engineers and Scientists*, 2nd ed., (Allyn & Bacon Inc., 1961), 424.

their initial phases. Then in the former case ($\omega_1=\omega_2=\omega$), the compound displacement x is (Fig. 1):

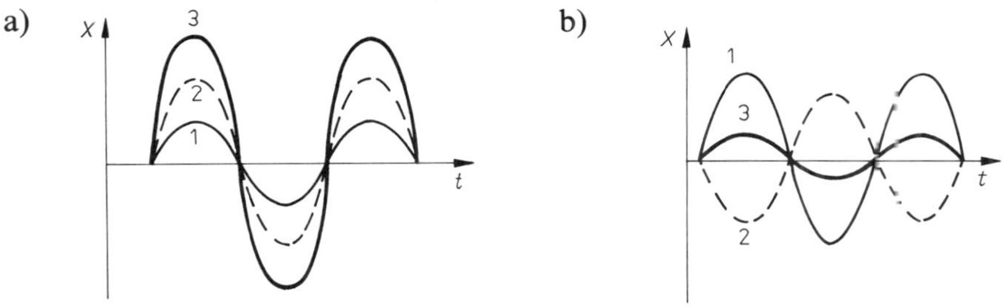

Fig. 1
An addition of two vibrations 1 and 2 of the same frequency: (a) the same phase, (b) the opposite phase; 3 is the compound vibration.

$$
\begin{aligned}
x_1 &= a_1 \cos(\omega t + \alpha_1) \\
x_2 &= a_2 \cos(\omega t + \alpha_2) \\
x &= x_1 + x_2; \\
x &= A \cos(\omega t + \alpha)
\end{aligned}
$$

where

$$A = \sqrt{a_1^2 + a_2^2 + 2a_1a_2 \cos(\alpha_2 - \alpha_1)} \tag{1.1}$$

In the latter case, assuming $a_1 = a_2 = a$ and $\alpha_1 = \alpha_2 = \alpha$; $|\omega_2 - \omega_1| \ll \min\{\omega_1, \omega_2\}$ we have (Fig. 2):

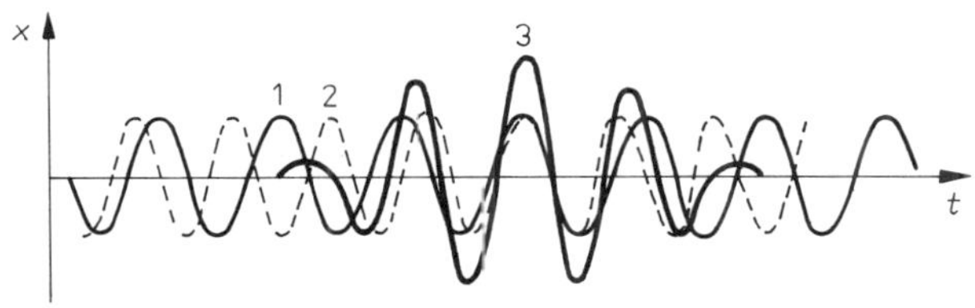

Fig. 2
An addition of two vibrations 1 and 2 of different frequencies, 3 is the compound vibration.

$$
\begin{aligned}
x &= a \cos(\omega_1 t + \alpha) + a \cos(\omega_2 t + \alpha) \\
&= 2a \cos\left(\frac{\omega_1 + \omega_2}{2}\right) t \cos\left(\frac{\omega_2 - \omega_1}{2}\right) t;
\end{aligned}
$$

or

$$x = A\cos\left(\frac{\omega_1 + \omega_2}{2}\right)t$$

where

$$A = 2a\cos\left(\frac{\omega_2 - \omega_1}{2}\right)t; \tag{1.2}$$

This means that the amplitude of compound vibration varies in time with the frequency $|\omega_2 - \omega_1|/2$. In both cases, the compound amplitude can be as high as double amplitude of each vibration and as low as zero.

A wave may be treated as a motion through a number of connected oscillators, each obeying the principle of superposition of vibrations. Thus, we may introduce the following *principle of superposition of waves:* when two waves intersect the displacement of the compound vibrations may be found by applying the principle of superposition of vibrations to every point of space, which may be achieved by adding the functions representing the two waves. For instance, let x_{01} and x_{02} are vibrations (similar to the ones described above), which occur at the centers C_1 and C_2 (Fig. 3). Let they travel from these centers with velocity V and

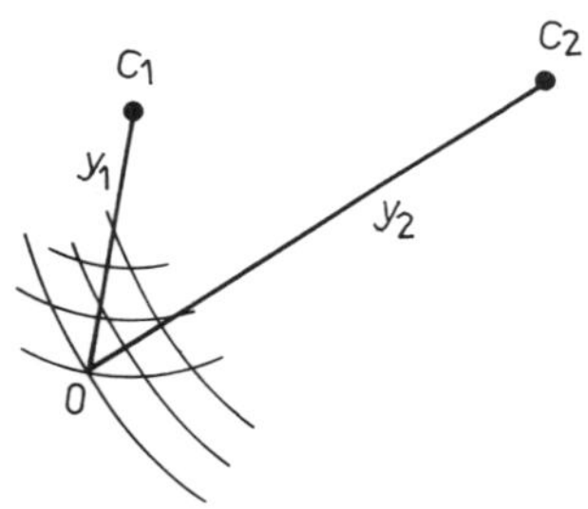

Fig. 3
Superposition of waves of the same frequency

reach the meeting point O, which is removed from the two centers by, respectively, y_1 and y_2. The result of superposition of these waves at the point O is ($P = y_1 - y_2$ is the path difference, T is the period, and λ is the wavelength):

$$x_{10} = a_1\cos(\omega t + \alpha);$$
$$x_{20} = a_2\cos(\omega t + \alpha);$$

At the point O the vibrations are:

$$x_1 = a_1 \cos\left(\omega t + \alpha - \frac{\omega y_1}{V}\right);$$

$$x_2 = a_2 \cos\left(\omega t + \alpha - \frac{\omega y_2}{V}\right); \qquad x = x_1 + x_2;$$

$$x = A \cos(\omega t + \alpha')$$

where

$$A = \sqrt{a_1^2 + a_2^2 + 2a_1a_2 \cos \frac{\omega(y_2 - y_1)}{V}}; \tag{1.3}$$

since $\frac{\omega}{V} = \frac{2\pi}{\lambda}$; the maxima will be observed where $\frac{2\pi P_{\max}}{\lambda} = m\pi$, and minima where $\frac{2\pi P_{\min}}{\lambda} = (2m-1)\frac{\pi}{2}$; thus,

$$P_{max} = 2m \cdot \frac{\lambda}{2}; \quad P_{\min} = (2m-1)\frac{\lambda}{2}; \quad m = 1, 2, 3 \ldots. \tag{1.4}$$

Thus, the intensity of compound vibrations (the intensity is proportional to the square of amplitude) may vary from zero to the quadruple intensity of each vibration (if both amplitudes are equal). The maxima or minima of intensity will occur at the points where the path difference of two waves equals, respectively, an even or an odd number of the half-wavelengths.

Based on superposition of vibrations, the principle of superposition of waves is applicable only to waves of small amplitude. In fact, this covers most phenomena known in Young's time. I will argue that this principle was discovered by Thomas Young.

If, within its range of application, the principle of "superposition" is simply a mathematical rule, the concept of "interference" will refer to *observable* physical phenomena, in which the perceived intensity of sound, light, or water waves experiences considerable periodical changes. (The perceived intensity is assumed to be proportional to the intensity of vibrations.) Hereafter, in addition to this general term I will use where it is necessary its two specific versions. One is the *hypothesis of interference*: the perceived periodical variation of intensity can be explained by a superposition of intersecting waves. The plausibility of this hypothesis for sound is supported by a periodical rise and fall of loudness which occurs when two tones of a slightly different frequency sound together ("beats of sound"). Thus, the hypothesis of interference implies a possibility of an "abnormal" addition of physical effects, such, for instance, as two sounds silencing one another. Since we assumed that all observed periodical variations of intensity can be explained through the hypothesis of interference, we may call them *the phenomena of interference*.[20]

While the hypothesis of interference provides a *qualitative* explanation of the interference phenomena, the other concept – *the principle of interference* – instructs how to apply the principle of superposition of waves to create a *quantitative* theory. There are several steps in building such a theory. First, we observe a given phenomenon and check that it is periodical. Since it is easier for our senses to notice the extreme magnitudes of the intensity than the intermediate ones, we concentrate on maxima and minima of intensity, which occur either in time or in space. (Naturally, we must observe more than one maximum or minimum.) Then we measure the intervals between neighboring maxima (minima); if they are constant, it is a true phenomenon of interference. The next step is to identify the waves, to which we will apply the principle of superposition of waves (for simplicity sake, we may begin with two waves). Then we formulate the rule either for the positions of maxima and minima or for intervals between them. This rule will depend on whether the intensity changes only in time *(temporal interference)* or only in space *(spatial interference)*. Spatial interference is produced by waves of the same frequency, the maxima or minima of spatial interference occur where the path difference of two waves equals, respectively, an even or an odd number of the half-wavelengths. In case of temporal interference, the waves have slightly different frequency, and the frequency of "beats" equals one half of the difference of the two frequencies. Finally, we compare the interval between the neighboring maxima (minima) calculated from the theory with the one measured experimentally, and, if necessary, we adjust the theory by adding new hypotheses.

While the principle of superposition is applicable to *any* two waves (within the range mentioned above), *not* every pair of intersecting waves interfere. I will call the waves which do interfere *coherent*, and the limitations set on them *the conditions of coherence*. The conditions of coherence are different for temporal and spatial interference and also depend on the type of waves and detector. For instance, as formulated in some optics texts, interference of light can be seen by the eye if the interfering waves have: 1) the same frequency ("the condition of frequency"); 2) a common origin ("the condition of origin"); 3) almost the same direction ("the condition of direction"); 4) a small path difference ("the condition of path difference"), and 5) come from a small luminous body ("the

[20] This is a very general term, applicable to interference of all kinds of waves. To the contrary, authors of books on optics use "interference phenomena" to designate a narrow class of the phenomena of interference of light, in which reflecting or refracting bodies are considered to be unlimited.

condition of the source's size"). In fact, there cannot be a "complete" list of these conditions, for some of them are interconnected and not all of them are necessary in every particular case.[21]

These limitations are caused by some properties of waves and their detectors. If our senses or instruments could respond instantaneously, *any* two intersecting waves would be coherent. In reality, however, due to its inertia a detector has some *resolution time* and cannot notice too rapid changes. Also, waves may experience irregular changes of amplitude or phase, which disrupts the interference pattern. Thus, the longer the time interval between two irregularities *(coherence time)*, the greater the chance to notice interference. For example, one can notice temporal interference only if its period is greater than the resolution time of the detector and smaller than the coherence time of waves. As to spatial interference, the coherence time of each wave must be much greater than the resolution time of the detector. In both cases, the length of observation must exceed the coherence time, and be greater for a quantitative investigation of interference than for a qualitative one.[22]

Thus, the *principle of interference* includes the hypothesis of interference, the rule for locating maxima and minima, the conditions of coherence, and two additional hypotheses. One is that interfering waves must be *periodical*, and the other is that one can account for all periodical colors by using only *two* interfering waves. This principle, unlike the principle of superposition of waves and the hypothesis of interference, is not universal. The rule for maxima and minima

---

[21] The modern theory of coherence is a mathematical theory, which deals with the function called "degree of coherence" (see, for instance, M. Françon, *Optical Interferometry* (Academic Press: New York & London, 1966), ch. I). Physically, this function describes the degree of visibility or contrast of fringes: the greater the degree of visibility the more fringes we can see. The conditions of coherence formulated above provide a qualitative approximation to this theory in the following sense: if the condition requires a parameter (for instance, the angle between interfering rays, or their path difference, or the angular size of the luminous body, or the bandwidth of the spectrum) to be small, it means the smaller this parameter the greater the degree of visibility and more fringes can be observed. The parameters are interconnected and we may change them so as to preserve interference even at unusual conditions. For instance one can obtain interference at a very large path difference by using light of a very high degree of monochromaticity. Or, in some cases, interference can be obtained with a large luminous source. This means that the conditions of coherence are *not* absolute and must be considered all together. This also mean that one may vary to some extent the set of conditions. For instance, one may use either the condition of direction or the condition of path difference; one may also introduce the condition of monochromaticity of light. There is also a special condition for polarized light.

[22] For instance, the ear cannot recognize more than about 10 beats of sound per second, and the coherence time of sound should be several seconds. Similarly, since the coherence time of light is 10^{-8} sec, the eye with its resolution time of about 0.1 sec cannot notice any "optical beats" and is limited to observing only spatial interference of light. However, a photomultiplier with a resolution time 10^{-10} sec can see the interference of light of different frequency, provided their wavelengths differ less than 10^{-2} nm (see Françon, 32–34).

differs for spatial and temporal interference, and the conditions of coherence depend on the type of waves and their detector.

To decide on when the hypothesis of interference and the principle of interference were accepted we have to agree on the meaning of the term "acceptance."

I.5 What is the "acceptance of a theory"?

Until now, historians have evaluated the response to the wave theory and the principle of interference by the view of the majority of scientists who responded to them. However, this approach does not help to understand either the initial rejection of the concept of interference or its subsequent acceptance. It seems that this can be achieved by examining a variety of opinions about the concept of interference and the reasons for their difference.

I found three factors that affected the responses to the concept of interference most prominently: a scientist's view on the nature of light, his philosophy of science, and his proficiency in physics (particularly in optics). Accordingly, I divided all commentators into three groups. The first ("quantitative physicists") consists of scientists who aimed at establishing mathematical laws of physical phenomena. The second group ("qualitative physicists") embraces those who were content with constructing qualitative mechanical models of optical phenomena. The third group ("non-experts") includes scientists who were not proficient in physics (chemists, biologists, physicians) and derived their opinions on the principle of interference from others.

We should also take into account the *level* of acceptance of a theory. Making an approving comment is not the same as promoting the theory in books and lectures, which in turn is different from applying the theory. Being concerned with the *development* of the principle of interference, I will be interested primarily in the higher levels of its acceptance. When I use the term *establishment* of a theory I mean that a theory is "established" when its status reaches that of other theories recognized at the time or, that the theory is *accepted* to the same degree or better than the competing theories.

With the necessary terms defined, we are now fully equipped for studying the discovery of the concept of interference.

Chapter II

Thomas Young and the problem of intersecting sounds

Section I: Young

II.1 What did Young discover about interference?

There is no agreement among historians about the meaning of Young's discovery of the concept of interference. Robert Silliman, for instance, stated that, "The novelty of Young's discovery lay not in the general concept of interference but in its application to optics. More than a century before, the idea of interference had been used by Newton himself . . . The idea also found an important application in acoustics." [1] The early Young scholars did not view this fact as diminishing the importance of his discovery. As John Herschel (1792–1871) wrote it 1830,

> The *principle of interference*, first distinctly referred to by Dr. Young as applicable to all motions of the vibratory or undulatory kind . . . cannot be considered as possessing much novelty. It is a particular case of a more general principle, that of the *superposition of small motions*, which had been admitted and employed by all mathematicians since Newton, in every enquiry where it could be of service, being, in fact, one of the *most obvious and elementary* of the consequences of the laws of motion. Newton himself had employed it . . . in the explanation of the curious affections of the tides at Batsham . . . That Dr. Young had read this passage and dwelt upon it, there is no reason to doubt, from his own repeated allusions to it; nor is it at all impossible, or in the slightest degree derogatory from his merit that by generalizing its terms and transferring its application, he might have obtained the first notion of its importance and availability in other cases. *In science, he who generalizes, invents*.[2]

[1] Silliman, 96.

[2] John Herschel to Hudson Gurney, June 25, 1830 in S.Schweber, ed. *Aspects of the Life and Thought of Sir John Frederick Herschel*, vol. 1 (New York, 1981), 148 (p.3 of the letter). Italics added.

Thus, Herschel envisioned the transition from mechanics to optics as a difficult one. To the contrary, in Worrall's view, "It was *not much of a step* to 'invent' the hypothesis that light waves interfere, given mechanics and given Newton's work both on periodicity and on interference."[3] Latchford sees his duty in destroying the "myth surrounding Thomas Young" by proving "Young's indebtedness to his predecessors for many of his own ideas."[4]

Thus, whatever their views on the value of a scientific generalization, all these authors agree that Young transferred to optics a mechanical principle, which he borrowed from his predecessors. They only differ on whether to call it "the principle of interference" or "the principle of superposition."

If the mechanical concept of interference was well known before Young and its transfer to optics was as simple as stated above, it is difficult to explain the total absence of competition in this discovery. Actually, the same argument is valid for Young's application of the concept of interference to sound and water waves: not only was there no one who challenged Young's priority in these two fields but no one even understood what he had achieved there.

I intend to show that Young's contribution to the discovery of the concept of interference as asserted above is underrated, and one of the reasons for this is a view that the concept of interference is a *simple and easy to grasp* idea. To clarify this subject it is necessary to answer the following questions: 1) what did scientists know about mechanical superposition and interference before Young? 2) what was Young's contribution, if any, to this field? and 3) what kind of difficulties did he encounter when attempting to extend this concept into optics? Latchford had already investigated the first two problems, however, unsatisfied with some of his solutions, I feel a need to reexamine them. In particular, I intend to demonstrate that the principle of superposition of waves and the principle of interference were unknown before Young. I will start this discussion with Young's early work on intersecting sounds and then move on to his predecessors.

II.2 Young on interference of sound

Young developed an interest in acoustics during his medical studies. While preparing his dissertation in 1796 in Göttingen, he was required to present a lecture on some topic connected with medicine, and he chose the formation of the human voice. After completing this research, he found himself so engrossed in

[3] Worrall, 137. Italics added.
[4] Latchford, 10, 11.

acoustics, that he continued to study it both experimentally and theoretically for the next three years at Cambridge. To get an English medical degree he had to satisfy residence requirements at an English university. Since Young felt no need to improve his medical education, he was able to spend this time as he chose. Young presented the results of his acoustical studies in his article "Outlines of Experiments and Inquiries Respecting Sound and Light," which he finished on July 8, 1799 and read to the Royal Society on January 16, 1800.[5] The paper revealed a fair knowledge of both English and Continental acoustical writers. Young frequently cites Lagrange, Euler, and Bernoulli; he also mentions Newton, Taylor, Maclaurin, Matthew Young, Chladni, Venturi, and others. The article appears to be a supplement to and a comment on eighteenth-century acoustical works, rather than an original research with a unified goal. He discusses the principal acoustical subjects of the time, adding to each a question, a clarification, or a new observation.

The section XI "Of the Coalescence of Musical Sounds" is known among historians for presenting the first application of the principle of interference to sound, sometimes called the "acoustical principle of interference." In fact, the section mentions no *new* principle and focuses instead on correcting Robert Smith:

> It is surprizing that so great a mathematician as Dr. Smith could have entertained for a moment, an idea that the vibrations constituting different sounds should be able to cross each other in all directions, without affecting the same individual particles of air by their joint forces: undoubtedly they cross, without disturbing each other's progress; but this can be no otherwise affected than by each particle's partaking of both motions. *If* this assertion stood in need of any proof, it might be amply furnished by the phenomena of beats, and of the grave harmonics observed by Romieu and Tartini; which M. De La Grange has already considered in the same point of view.[6]

Here Young presents the reader with one of the major problems of the eighteenth-century acoustics: how sound waves cross without disturbing one another. To him it is obvious that this problem cannot be solved without using the concept of superposition. (As we will see further, Smith was not the only one

[5] Young, "Outlines of experiments and inquiries respecting sound and light [1799]," *Phil. Trans.* 1800: 106–50. The references arre given to *Misc. Works*, 1: 64–98. Hereafter, this article will be referred to as "Outlines."

[6] Young, "Outlines," 83, italics added

guilty of rejecting this solution: its advantages were not as obvious to others as they were to Young.)

In the next paragraphs Young illustrates how the phenomena of beats and grave harmonics support the concept of superposition.[7] First, he shows how to superpose two vibrations graphically:

> ... let us suppose ... that the particles of air, in transmitting the pulses, proceed and return with uniform motions; and in order to represent their position to the eye, let the uniform progress of time be represented by the increase of the absciss, and the distance of the particle from its original position, by the ordinate, Fig. 93–98 [see Fig. 4] Then by supposing any two or more vibrations in the same direction to be combined, the join motion will be represented by the sum or difference of the ordinates.[8]

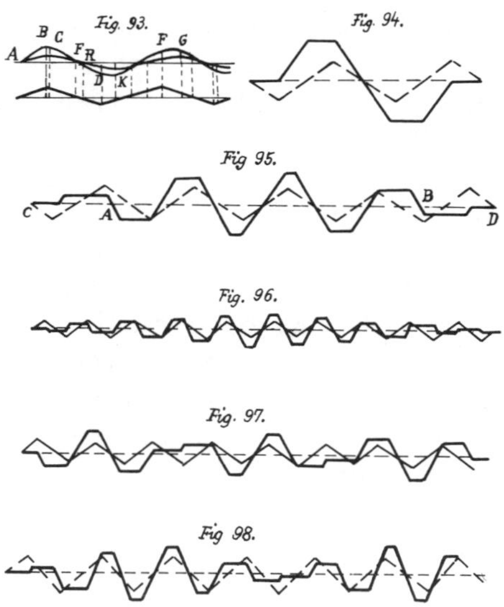

Fig. 4
Young's illustration of superposition of sound waves of different frequencies (from Young, *Misc. Works* 1: plate III).

[7] The grave harmonic is a perception of a third tone, produced by two strong tones of different frequency, which sound simultaneously. It was described in the middle of the eighteenth century by Sorge, Tartini, and Romieu as a sound lower than the original tones, which explains its name. From their descriptions Lagrange derived that the frequency of this grave harmonic was equal to the difference of the two frequencies. Much later, this sound was named the "differential tone."

[8] Young, "Outlines," 83.

This passage together with the related diagrams presents a mathematical principle, which may be interpreted in our terms as the principle of superposition of vibrations. Since in case of vibrations of different frequencies the diagrams for a superposition of vibrations and for a superposition of waves look the same, we may say that we see here the first display of the principle of superposition of waves.

In the next passage, Young passes from a mathematical principle to phenomena:

> When two sounds are of equal strength, and nearly of the same pitch, as in Fig. 96, the joint vibration is alternately very weak and very strong, producing the effect denominated a beat . . . The greater the difference in the pitch of two sounds, the more rapid the beats, till at last . . . they communicate the idea of a continued sound; and this is the fundamental harmonic described by Tartini. For instance, in Figs. 94–97, the vibrations of sound related as 1:2, 4:5, 9:10, and 5:8, are represented; where the beats, if the sounds be not taken too grave, constitute a distinct sound, which corresponds with the time elapsing between two successive coincidences, or near approaches to coincidence.[9]

Since the explanation of these phenomena involves a superposition of two waves and the conditions of coherence (equal amplitude and almost the same frequency), this is the principle of interference of sound (more exactly, of temporal interference).

It is clear that when speaking of an intersection of sound waves Young distinguishes the mathematical procedure (an addition of two functions) from the observable physical effect, which he calls "coalescence." He does not discriminate, however, between different types of superposition. For the reason explained above, this did not affect Young's explanation of the beats and the grave harmonics. However, as we will see further, it hurt him badly in another way.

When comparing other sections of the "Outlines", rich in new experiments or ideas, with the section XI, one wonders whether the latter had any other goal but refuting Smith. The following passage, missed by historians, shows there was one:

> But, besides this primary harmonic, a secondary note is sometimes heard, where the intermediate compound vibrations occur at a certain interval, though interruptedly; for instance, in the coalescence of two

[9] Ibid., 83–84.

sounds related to each other as 7:8, 5:7, or 4:5, there is a recurrence of a similar state of the joint motion, nearly at the interval of $\frac{5}{15}$, $\frac{4}{12}$, or $\frac{3}{9}$ of the whole period: hence in the concord of a major third, the forth below the key note is heard as distinctly as the double octave, as it seen in some degree in Fig. 95; AB being nearly two thirds of CD.[10]

Young demonstrates here that two tones sounding together may produce more than one third sound. In this he anticipates in part the discovery of combination tones by Hermann Helmholtz (1821–1894).[11] It is important to note that, unlike the differential tone, Young's grave harmonics could not be explained without his principle of superposition of waves (see II.3).[12]

Thus, Young is using here some new ideas (in our terms, the principle of superposition of waves and the principle of interference) without claiming any prior-

Fig. 5
A reconstruction of the addition of two vibrations in Young's Fig. 95 (Fig. 4).

[10] Ibid., 84.

[11] Helmholtz discovered that two tones of different frequencies ω_1 and ω_2 when sounding together produce a number of "combination tones" with frequencies $|n\omega_1 \pm m\omega_2|$ where n and m are integers. If in the example given by Young $\omega_1 = 400$ Hz, and $\omega_2 = 500$ Hz, then the frequency of his combination tone was $\omega = 300$ Hz. This may be interpreted as $\omega = 2\,\omega_1 - \omega_2$, which means that Young's tone was not the differential one. Nonetheless, when referring to it I will retain Young's term "grave harmonic."

[12] Let us examine the diagram in Young's Fig. 95 (our Fig. 4). The bold curve is the result of adding of two triangular curves, only one of which is present in the diagram. My Fig. 5. shows the addition of two triangular vibrations: I preserved the same ratio of frequencies as in Young's Fig. 95 but not the amplitude, and as a result the compound vibration in Fig. 5 somewhat differs from that in Fig. 95. Horizontal tops in Young's diagram indicate that he selected two triangular functions with the same angle. (In my dissertation, I had assumed that both curves in Young's Figs. 95–98 were the *primary* vibrations, but later I realized the mistake.)
In Fig. 95 the length *CD* represents the period of the compound curve, and since it contains four periods of the lower component, the Tartini grave harmonic (the "differential tone") is a double octave below it. To explain the appearance of another grave harmonics Young is looking for a regular part *AB* within the quite irregular compound curve *CD*. He supposes that a periodical repetition of this part will be perceived as a musical tone, despite the irregular parts *CA* and *BD*. Since *AB* contains three periods against four periods of the lower component, Young's grave harmonics makes with this component a major third (the ratio of frequencies is 3:4). As shown in II.4, the representation of sound, which was applied to explain the beats before Young, was too simple to account for Young's grave harmonic.

ity for them. It appears he doubted that these ideas were new enough. This transpires from a comparison of several of Young's comments about his acoustical discoveries. I will review them in chronological order.

In 1798, he thought they were new:

> I am ashamed to find how much the foreign mathematicians for these forty years have surpassed the English in the higher branches of the sciences. Euler, Bernoulli, and D'Alembert have given solutions of problems which have scarcely occured to us in this country. I have had particular occasion to observe this in considering the figure of vibrating chords, the sounds of musical pipes, and some other similar matters, in which I fancied I had hit on some ideas entirely new, but I was glad to find them in part anticipated by Bernoulli in 1753 and 1762. There are *still* several particulars respecting the gyration of chords, the formation of synchronous harmonics, the combination of sounds in the air, the phenomena of beats, on which I flatter myself that I shall be able to throw some *new light*, and to correct several misstatements of Dr. Smith.[13]

Young does not mention here the grave harmonics (possibly he is still unaware of this phenomenon), and he believes that his explanation of beats is new. It is possible that he dropped this claim in the "Outlines" because he had read Lagrange's paper some time between July 1798 and July 1799. Whatever the reason, soon Young changed his mind. In his letter to William Nicholson, of July 13, 1801, Young lists the application of the principle of superposition of waves to sound and the discovery of two grave harmonics among his major results in the "Outlines" (the explanation of beats is not mentioned).[14] In 1802, Young says: "I was not aware that . . . there was *so much novelty* in the mode of obtaining them, as to deserve the name of a *theory* or an *invention*."[15] Thus, he admits *some* novelty. He retains a similar opinion in 1804, when listing the phenomena of beats and of grave harmonics among the subjects, which were examined "in a *new point of view*."[16] However, in 1807, Young stops claiming any innovation, and presents the superposition of sound waves as a consequence of "general laws of

[13] Young to Andrew Dalzel, July 8, 1798, in Dalzel, *History of the University of Edinburgh from its Foundation*, 2 vols (Edinburgh, 1862), 1:16[illegible], italics added.

[14] Young. "A letter to Mr. Nicholson . . . respecting sound and light, and in reply to some observations of Professor Robison [July 13, 1801]," *Misc. Works* 1:133.

[15] Young, "An answer to Mr. Gough's 'Essay on the theory of compound sounds'," *Nich. Jour.* 2 (August 1802): 264, italics added.

[16] Young, "Reply," 200. Italics added.

composition of motion, which are the foundation of the principal doctrines of mechanics." [17]

This Young's vacillation indicates that he believed he had been anticipated in this field but was not certain to what extent. According to Latchford, the section XI "represents Young's first public enunciation of the super-position principle as it applied to sound (probably the first to be made in England)," [18] and "many writers before Young recognised super-position." [19] The implication is that Young inherited the principle of superposition from his predecessors and added nothing to it.

Let us now examine what could Young have borrowed from the eighteenth-century acoustics. I will group the relevant works according to their concern either with the interaction of waves or with their independence.

Section II: Young's predecessors

II.3 Reinforcement of sound

Long before Young, scientists knew that two musical tones sounding together may create new effects, such as consonance and dissonance, beats of sound, and some others, and explained them by combining actions of the two sounds. First, it was done for consonance: "it has been received for an undoubted Axiom, that *Consonance* is made by the frequent union of two Sounds in striking the External Drum of the Ear . . . at one and the same time." [20] In a similar way Joseph Sauveur (1653–1716), a member of the Paris Academy of Sciences, explained beats of sound in 1700: "the sound of two organ-pipes sounding together must have more force when their vibrations, after being separated for some time, come to unite and accord to strike the ear by the same blow." [21] It is important to note that Sau-

---

[17] Young, *A Course of Lectures on Natural Philosophy and Mechanical Arts*, 2 vols (London, 1807), 1: 389. Hereafter, this book will be referred to as *Lectures*.

[18] Latchford, 132.

[19] Ibid., 3.

[20] "An Account of *Musica speculativa* del Mengoli," *Phil. Trans.* 8 (1674): 6194.

[21] "le son de deux tuyaux ensemble devoit avoir plus de force, quand leurs vibrations, après avoir été quelque temps séparées, venoient à se réunir, & s'accordoient à frapper l'oreille d'un même coup." See "Sur la détermination du son fixe," *Hist. Acad. Paris*, 1700: H171; hereafter cited as "Du son fixe." Each volume of this periodical contains "Histoires" and "Mémoires" (the pages will be marked, accordingly, "H" and "M"). "Histoires" consist of the abstracts of papers, prepared probably by Fontenelle, the Secretary of the Academy. Latchford interprets this passage as stating that sounds add *outside* the ear.

veur represented sound here by discontinuous pulses and not by sinusoidal waves as physicists usually did. At least, that is how Robert Smith (1689–1768), the Plumian Professor of Astronomy at Cambridge and the Master of Trinity College, understood him: "though the pulses of sounds of a different pitch have different durations, they may yet be *abstractly considered* as if they were instantaneous; by taking only the middle *instant* of each pulse."[22] Such model cannot be applied for subtracting the actions of two sounds, which implies that the decrease of the original loudness was overlooked.

In Smith's view, Sauveur's explanation of consonances and beats is fallacious. According to Sauveur, two tones, which beat slowly (i.e. coincide infrequently), produce a *discord*, while fast beats make an *accord*.[23] But, Smith says, in a tempered system the harmony is perfect although the frequencies of sounds are incommensurable and thus their beats are extremely slow (practically, these sounds do not coincide at all).[24] To strengthen his point Smith offers another argument of the same type, in which he considers two sounds, the pulses of which bisect the intervals of one another (in modern terms, the sounds are of the same frequency but of the opposite phase):

> When any string of a violin . . . is moved by a gentle uniform bow, while its middle point being lightly touched by the finger, is kept at rest, but not pressed to the fingerboard; the two halves of the string will sound perfect unisons, an eighth above the sound of the whole; and will keep moving constantly opposite ways . . . Hence, because these opposite motions of the halves of the string communicate and propagate the like motions to the contiguous particles of air and these to the next successively, it follows that different particles of air at the ear, placed any where in a perpendicular that bisects the whole string, will keep moving constantly opposite ways at the same time; *those* particles, which received their motion from one half of the string, going towards the ear, while *others* are returning from it, which received an antecedent motion from the other half of the string: Or, in fewer words, the successive pulses of one sound are constantly bisecting the intervals between the pulses of the other: and yet the harmony is perfectly agreable to the ear.[25]

[22] Smith, *Harmonics, or the Philosophy of Musical Sounds*, 2nd ed., (London, 1759), 8. Here, as well as in many other contemporary sources, the "pulse" stood for a vibration or a wave, not necessarily discontinuous. Italics added.

[23] Sauveur, "Du son fixe," H177. Italics added.

[24] Smith, 99.

[25] Ibid., 105. Italics added.

Thus, a consonance can be produced by sounds, the pulses of which never coincide: a strong blow to Sauveur's theory.

Finally, Smith comes to the beats: "it is very unreasonable to suppose with Mr. Sauveur that the beats are made by the *united force* of the coincident pulses of imperfect unisons."[26] His own theory of beats is obscure physically but mathematically it is identical with Sauveur's. To illustrate his point, Smith represents two sounds by two parallel rows of equidistant points (the ratio of intervals between the next points in the two rows equals the ratio of frequencies): beats occur where the two rows coincide (Fig. 6).[27]

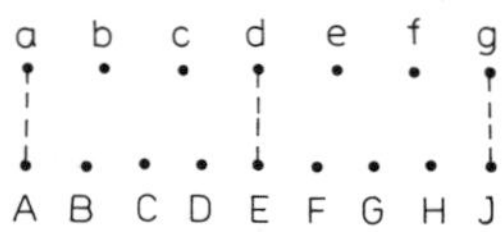

Fig. 6
An illustration of the beats of sound (from R. Smith, *Harmonics*)

Matthew Young (1750–1800), Professor of Natural Philosophy in Dublin,[28] and John Robison (1739–1805), Professor of Natural Philosophy in Edinburgh,[29] followed Sauveur's idea to account for a reinforcement of sound in a speaking trumpet. They suggested that in a trumpet of a proper shape pulses of sound, reflected by its walls, reach the ear simultaneously with direct pulses and unite their actions. This idea, though, was never developed into a quantitative theory.

Interestingly, in the phenomena described above physicists were looking only for a reinforcement of sound. In a similar way, in other cases they focused exclusively on its destruction.

[26] Ibid., 247.

[27] Ibid., 81–2. See also Plate XI.

[28] M. Young, *An Enquiry into Principal Phenomena of Sounds and Musical Strings* (London, 1784), 53, 60–61; hereafter is referred to as M. Young.

[29] J. Robison, "Trumpet" in *Encyclopaedia Britannica*, 3rd ed, 18 vols (Edinburgh, 1797). References are given to the reprint in J. Robison, *System of Mechanical Philosophy*, 4 vols (London, 1822), 4: 458, hereafter is cited as "Trumpet."

II.4 Destruction of sound

In 1673, William Noble and Thomas Pigot discovered an immobile point, or a *node*, in the middle of a vibrating string excited by another string half of its length.[30] In 1762, Daniel Bernoulli (1700–1782), demonstrated that nodes (or rather nodal planes) also exist in sounding pipes.[31] Finally, in 1787, a musician Ernst Florens Friedrich Chladni (1756–1827) discovered nodal lines in sounding membranes.[32] According to Bernoulli, a node was produced at such a point of an organ-pipe where two adjacent layers of air acted with the same force but in the opposite direction.[33] M. Young applied the same idea to explain the experiment of Noble and Pigot as follows. Since the period of vibrations of the short string *AB* is one half of that of the long one *CD*, there are moments when they move in the opposite directions and communicate to the air near the midpoint *E* equal and oppositely directed momenta (Fig. 7). With a stopped midpoint, the string *CD* will produce its second harmonic instead of the fundamental.[34]

This contradicted the opinion of Jean le Rond D'Alembert (1717–1783) that the two halves of the string *CD* will be as often supported as opposed by vibrations of the string *AB*, and thus the second harmonic will be destroyed.[35] The

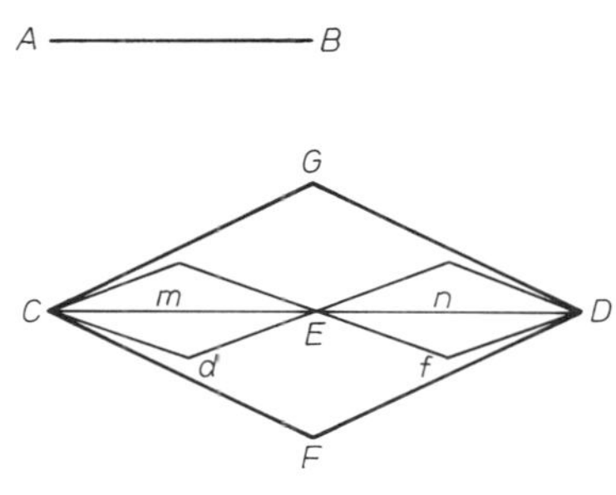

Fig. 7
An illustration of the production of a node (from M. Young, *Enquiry*)

[30] Wallis, "Letter to the Publisher, Concerning a New Musical Discovery," *Phil. Trans.*, 12 (1677): 839–41.
[31] D. Bernoulli, "Recherches physiques, mécaniques et analytiques, sur le son & sur les tons des tuyaux d'orgues differemment construit," *Hist. Acad. Berlin*, 1762: H170–81, M431–85.
[32] Chladni, *Entdeckungen über die Theorie des Klanges* (Leipzig, 1787).
[33] Bernoulli, "Recherches physiques," 433–4.
[34] M. Young, 88–91.
[35] Alembert, "Fondamental," in *Encyclopédie ou Dictionnaire raisonné des sciences, des arts et des métiers*, 17 vols (Paris, 1751–65), 7: 56.

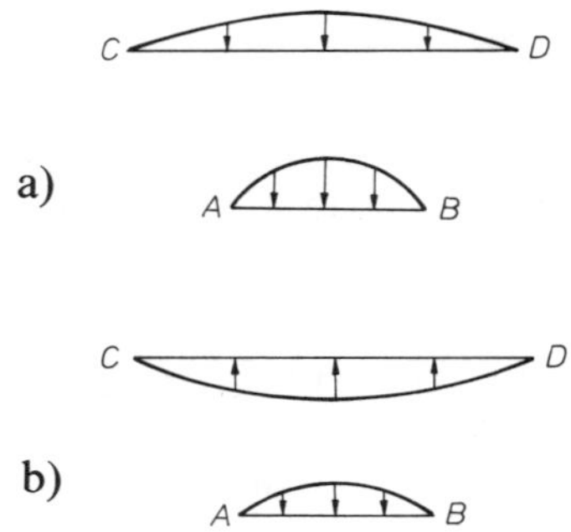

Fig. 8
An illustration of the dispute between D'Alembert and M. Young on whether two strings AB and CD support the vibrations of one another or destroy them: the former emphasized the situation (a), while the latter focused on the situation (b).

contradiction resulted from adding the motions communicated by the two strings at different moments (see the illustrations in my Fig. 8).

Since in all the phenomena discussed above two motions were added (or subtracted) only at specific moments of time, the principle applied there was that of the superposition of motions. The controversy between M. Young and D'Alembert shows that this principle is inapplicable to vibrations, however that was not realized at the time.

The authors of some explanations of resonating strings are speaking of concurring and opposing vibrations. Jean-Jacques D'Ortous de Mairan (1678–1771), a member of the Paris Academy of Science, says that a string set in motion by another string will sound only if its vibrations are supported by the pulses of air created by the second string, which occur when both strings are tuned in unison. If they are not, at some moments the vibrations of the second string will be opposed to by those of the first string, and its sound will be extinguished.[36]

M. Young specifies that a string can resonate to any other string of a commensurable length, since their vibrations will periodically support one another. However, if the strings are incommensurable, then every successive pulse of one will partly counteract the motion generated in the other by the preceding pulse.[37] He also occasionally speaks of waves traveling back and forth within a string, which "cross and break" one another.[38]

[36] Mairan, "Discours sur la propagation du son dans les differens tons qui le modifient," *Hist. Acad. Paris*, 1737: H136. Hereafter is referred to as Mairan.
[37] M. Young, 87.
[38] Ibid., 98, 100.

Although these statements resemble modern applications of the principles of superposition of vibrations and superposition of waves, the appearance is deceptive, for in no case their authors compare the results of the addition of two motions at *different moments* or at *different points*. What they actually use is the explanation of resonance known since the sixteenth century: a periodical support of vibrations increases their amplitude until one hears a sound, while lack of such support prevents the production of sound.

Thus, the mechanical principle applied before Young to explain reinforcement or destruction of sound was the principle of superposition of motions. Applying it instead of the principles of superposition of vibrations or superposition of waves was bound to lead to errors, such as the one committed by D'Alembert and M. Young. What could had been the causes which prevented scientists from discovering the principles of superposition of vibrations and waves? Perhaps, one was the tendency to see in each vibratory phenomenon either an addition or subtraction of motions but not both playing some role. Another was skepticism about the idea of sharing several vibrations by the same particle.

II.5 Do sound waves interact?

It had been presumed that sound waves produced by people speaking in the same room cross one another. Since experience showed that one can hear each person in a company so as if this person were alone, the conclusion was made that intersecting sound waves do not affect each other.

To explain this independence of sound scientists resorted to analogies. The most popular one was the analogy between sound and waves on the surface of water, which pass through one another without any change. Some authors also compared sound to light: light rays cross without disturbing each other because one could see through a small hole several objects of the same shape and color as if each of them were present alone.[39] To the "emissionists," this independence of light resulted from large intervals between light particles: one ray passed through the interstices between the particles in the other ray.[40] The "undulationists" stated that a particle, excited by several waves simultaneously, can transmit them without a change.[41]

[39] J. Lambert, "Sur quelques instrumens acoustiques," *Hist. Acad. Berlin* 19 (1763): 115.

[40] See, for instance, R. Haüy, *Traité élémentaire de physique*, 2 vols (Paris, 1803), 2: 150.

[41] Huygens, *Traité de la lumière* (Leiden, 1690); references are given to the English translation *Treatise on Light* (University of Chicago Press, [1912]), 17–18, 22.

Not all physicists approved these analogies. In Robison's view, one should not demonstrate independence of sound through independence of light, because the latter is also little understood.[42] Mairan objects to the comparison of sound to water waves on the ground that these waves have different origin (elasticity and gravity) and different properties: for instance, the velocity of sound does not depend on frequency, while this is not true for water waves.[43] To clarify Mairan's point Fontenelle presented the following argument of his own. When two waves are excited in water, he notes, a hand placed in water senses an "average movement," the direction and velocity of which depend on those of both waves; on the contrary, the ear perceives *each* of the intersecting sounds and not the "average" one (that was written before the discovery of the third sound).[44]

Mairan explains independence of sound without any analogies. He adopts a popular then theory of hearing, according to which the internal ear contains a great number of tiny fibers of different length, each of which resonates to a specific tone.[45] It had been assumed that when one hears a tone, there must be outside the ear a source of waves of the corresponding frequency. Similarly, an accord of three tones meant three waves of different frequencies entering the ear. If all these waves, Mairan argues, are transmitted by the same particle of air, the particle must take part simultaneously in three different vibrations and thus possess three different velocities. Thus, he concludes, sharing of motions is mechanically impossible.[46]

The only alternative left, in Mairan's view, is to postulate that different sounds are transmitted by different particles. (In 1670, Mengoli expressed an idea, worded similarly to Mairan's, although the passage is too obscure to be certain about its meaning.[47]) Mairan supposes that, due to a difference in elasticity or size, air particles have different proper frequencies of vibrations. Although any wave excites a vibration in a particle, its amplitude is much greater when the frequency of the wave coincides with the proper frequency of the particle. Thus, a particle, affected by many vibrations at the same time, will transmit only one of them.[48]

Although Smith does not refer to Mairan, he shares his theory. This is evident from the above cited passage about sounds bisecting intervals of one another.[49] If

---

[42] Robison, "Trumpet," 477.
[43] Mairan, M6–8.
[44] Mairan, H137.
[45] Ibid., M12–14.
[46] Ibid., M5.
[47] See n.20 above, 6196.
[48] Mairan, M9–12.
[49] See n.25 above.

different waves were transmitted by different particles, could they distort one another? Smith answers that such a distortion may result only from collisions of particles, which do not occur, however, for:

> in so rare a fluid as air is, where the intervals of the particles are 8 or 9 times greater than their diameters, there seems to be room enough for such opposite motions without impediment: especially as we see the like motions are really performed in water, which in an equal space contains 8 or 9 hundred times as many such particles as air does. For when it rains upon stagnating water, the circular waves propagated from different centers, appear to intersect and pass through or over each other, even in opposite directions, without any visible alteration in their circular *figure*, and therefore without any sensible alteration of their motions.[50]

What Smith is apparently trying to say is that one can easily explain independence of waves by assuming that different waves are transmitted by different particles. Indeed, in such a case waves could be distorted only because of the collision of particles; and if an intersection of water waves reveals no sign of collisions, this must be even more true for sound in air, since particles of air are further removed from one another than particles of water.

It was this passage that provoked Young's angry rebuke (n. 6). Young is more clear about this in his notebook: "Dr. Smith has made a very great mistake with respect to this composition of sound; for he imagines that one sound is transmitted by a different series of particles from another."[51]

Despite the criticism of Mairan and Smith, the idea of sharing motions was revived in the 1750s, this time by mathematicians.

II.6 Mathematical approach to independence of sound

Newton gave the first theory of sound in his *Principia*. He assumed that air particles interact according to Hooke's law and showed that a vibrating body set them into harmonic motion. In 1715, Brook Taylor discovered a similar law for a mo-

[50] Smith, 106. According to Smith, the interval between water particles is about nine times smaller than that between air particles ($900^{1/3}$), or comparable with their diameter. Thus, if water particles do not collide, although they almost touch one another, this will not happen to widely spaced air particles.

[51] T. Young, lecture notebooks, University College, London. MS Add 13/14, 15/3r.

tion of a vibrating string. During the 1740s and the 1750s, Bernoulli, D'Alembert, Euler, and Joseph-Louis Lagrange (1736–1813) criticised the sinusoidal solution of the wave equation as too specific and tried to find a general one. Lagrange and Euler also wanted to show that their new mathematical techinique was the only means to explain the independence of sound waves.

To achieve this Lagrange considered an air particle C_1 in an organ pipe AB, vibrating with the maximal velocity u_1, and examined the change of the velocity of a short pulse propagating from C_1 along the pipe.[52] When the pulse reaches another particle D it communicates to it the same velocity u_1, and the speed of the pulse does not depend on the velocity of vibrations. If there are several oscillators $C_1, C_2, C_3 \ldots$ vibrating with maximal velocities $u_1, u_2, u_3 \ldots$, respectively, the pulses coming from these points to D give this point the velocity of vibrations $u_1 + u_2 + u_3 + \ldots$ This means that presence of other pulses does not change the contribution of the first pulse. Moreover, Lagrange showed that the contribution of each pulse will be the same even if it comes to D after a number of reflections at the ends of the pipe, which implies that an intersection of pulses changes neither their velocities of vibrations nor the speed of sound. Euler confirmed Lagrange's results using a different mathematical technique.[53]

Thus, to Lagrange and Euler, a demonstration of independence of sound consisted of proving that individual pulses preserve their speed of propagation in the presence of other pulses, which was a mathematical problem never attacked before. If a particle is affected by two identical waves traveling in the opposite directions, it transmits them further without any changes. Although this case was an obvious consequence of his general theory, Euler found it necessary to discuss it separately because "this is exactly the case where one must think that the perturbations would confound one another."[54] Apparently, he referred to the view that if two identical balls, moving in the opposite direction with the same speed, stop one another, so could two waves. Thus, one of the basics of the wave motion – a possibility for a particle to transmit simultaneously several motions in different directions – was little understood at Euler's time. Huygens had explained it in his *Traité de la lumière*, but his book was forgotten.

Incidentally, with all its merits, the proof of independence of sound by Euler and Lagrange is circuitous. Indeed, its idea is that a sum of several solutions of a particular wave equation is a solution of the same equation, which means that an intersection of waves does not change their speed. However, the wave equation is

---

[52] Lagrange, "Recherches sur la nature et la propagation du son [1759]," in his *Oeuvres*, 14 vols (Paris, 1867–92), 1 : 139–42. Hereafter is cited as Lagrange.

[53] Euler, "De la propagation du son [1759]," *Opera omnia*, 3rd ser.3, v.1 (Leipzig, 1926), 444–50.

[54] Ibid., 449.

derived in the assumption of infinitely small vibrations, and its applicability to real waves must be established experimentally. Since non-linear effects were unknown in the eighteenth century, this defect of the proof had not been noticed and the demonstration was approved, at least by the adherents of mathematical physics.

This does not support, however, Latchford's view that the Lagrange-Euler theory could have helped in discovering the principle of superposition of waves.[55] Indeed, how much information about the amplitude of the compound motion can one obtain when adding two waves expressed by Euler's functions

$$y = \varrho(x + vt) + \psi(x - vt) \qquad (2.1)$$

where x is the coordinate, t is time, v is the velocity of sound, and ϱ, ψ are continuous functions?

Perhaps, it were such general functions that John Robison had in mind when saying that "we cannot (at least *no mathematician has yet done it*) discriminate and then combine these agitations, with the intelligence and precision that are necessary for enabling us to say what is the ultimate accumulated effect."[56]

II.7 Harmonics

Another stimulus for studying the independence of sound came from the phenomenon of harmonics. Musicians knew that a fundamental tone of a string or a wind instrument is sometimes accompanied by higher harmonics such as its octave, twelfth, fifteenth, or seventeenth. The source of these harmonics was sought in the string, in the air, and in the ear.

Rene Descartes (1596–1650) thought that harmonics appeared in the instrument and were produced by parts of a string vibrating independently of the string as a whole.[57] Later, Sauveur specified that the octave was produced by one half of the string, the fifteenth – by one third, and so on.[58] Bernoulli accepted this hypothesis as an experimental fact and used it to solve the problem of vibrating

[55] Latchford, 101.
[56] Robison, "Trumpet," 477. Italics added.
[57] "Descartes to Mersenne, July 22, 1633," in *Correspondance du P. Marin Mersenne*, 14 vols (Paris, 1945–1980), 3: 458.
[58] Sauveur, "Systeme génerale des intervalles des sons, & son application à tous les systemes et à tous les instruments des musique," *Hist. Acad. Paris* 1701: M463.

string.[59] He supposed that a vibration of a string may be given by a sum of its harmonics:

$$Y = \sum_{m=1}^{\infty} \alpha_m \frac{m\pi x}{l}; \tag{2.2}$$

where Y is the initial shape of the string, m is the number of the m_{th} harmonic, α_m is its amplitude, x is the coordinate of a point, and l is the string's length. Mathematically, this equation means that any "continuous" function may be approximated by a trigonometric series.[60] Physically, it states that any excitation of a string creates simultaneously with its fundamental tone a number of harmonics, although in particular cases some of them may be missing.

To illustrate how a compound motion of a string can be obtained graphically, Bernoulli added the first two harmonics (Fig. 9). Young knew Bernoulli's paper

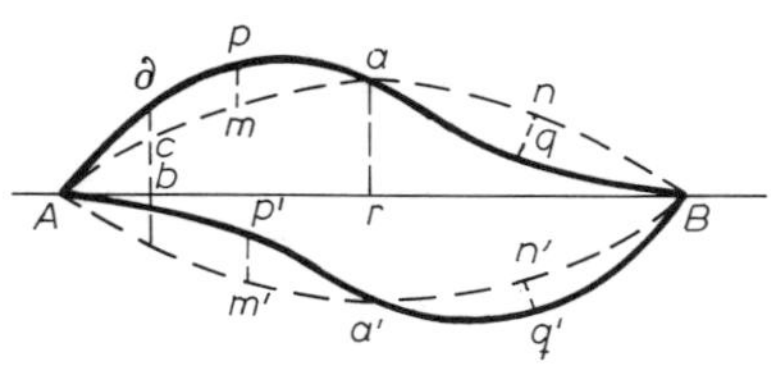

Fig. 9
Bernoulli's method of adding harmonics: ApaqB is the result of the addition of the fundamental AmanB and its second harmonic (from Bernoulli, "Reflexions," Fig. 6)

and probably borrowed from it the idea of adding vibrations graphically. However, his purpose and procedure were different from Bernoulli's. Young added two curves to find how the compound amplitude changes with time, while Bernoulli wanted to express the concept of a relative vibration, or a vibration relative the point which in its turn participates in another vibration. For instance, to add the first two harmonics, he drew the second harmonic *AdpaqB* so that its axis would coincide with the first harmonic *AmanB*. The most important, however, was that Young added *arbitrary* periodical curves, while Bernoulli dealt exclu-

59 Bernoulli, "Reflexions et éclaircissemens sur les nouvelles vibrations des cordes exposées dans les mémoirs de l'Académie," *Hist. Acad. Berlin*, 9 (1753): 147–72.

60 On the connection between mechanics and the problem of mathematical continuity see C. Truesdell, *The Rational Mechanics of Flexible or Elastic Bodies, 1638–1788*, in *Leonhardi Euleri Opera Omnia*, 2nd ser., v. 11 (Turin, 1960), 243–44.

sively with sinusoids of multiple frequencies. Since a superposition of such curves does not reveal any easily recognizable pattern of the amplitude change, Bernoulli hardly had an opportunity to discover the hypothesis of interference.

In 1762, Bernoulli offered a new experiment to demonstrate his hypothesis.[61] A rotating cog-wheel periodically striked a vibrating string. By changing its speed, Bernoulli was able to produce different harmonics in the string. Later, M. Young performed a similar experiment. While Bernoulli did not claim to produce different waves *simultaneously*, M. Young did: "I observed two forms of a chord, one vibrating very rapidly, while, *at the same time*, the other appeared to roll slowly backwards and forwards, and to cross the former." [62] He did not know, of course, that the effect was due to a peculiar feature of perception: to switch attention from one mode to another.

The hypothesis of coexistence of different vibrations had a number of opponents. Mairan noted that every point of a vibrating string belongs to many different aliquot parts and thus it must have many different velocities at the same time, which is impossible mechanically. His other argument was that harmonics cannot exist in a vibrating string without producing nodes, however an unloaded vibrating string never displays any nodes. Mairan concluded that the source of harmonics is not in a string but in the air. He supposed that a sound of frequency f excites not only the particles with the proper frequency f but also those with proper frequencies $2f$, $3f$, and so on. The waves emitted by the latter particles will be perceived by the ear as harmonics.[63] D'Alembert asserted that even if harmonics somehow appeared in a vibrating string their vibrations would have been frequently opposed by the fundamental and eventually destroyed.[64] Lagrange agreed with Mairan's criticism but not with his solution of the problem and suggested that harmonics were produced not in the vibrating string but in the surronding bodies, which resonate to the string.[65] This view won support of Euler.[66]

Jean-Adam Serre (1704–1788), a Swiss scientist and a musical theorist, was the only one who suggested that harmonics originate in the ear.[67] He assumed that each sound wave of frequency f could affect several fibers, namely those which were tuned to frequencies f, $3f/2$, $2f$, $3f$, and so on. This explained why harmonics appear together with the fundamental tone.

---

[61] Bernoulli, "Recherches physiques," 443.

[62] M. Young, 77–78. Italics added.

[63] Mairan, 16–18.

[64] D'Alembert, "Fondamental," 55–56.

[65] Lagrange, 147.

[66] Euler to Lagrange, 23 October, 1759. *Correspondance de Leonhard Euler* (Birkhäuser Verlag: Basel, 1980), 425.

[67] J. Serre, *Observations sur les principes de l'harmonie* (Geneve, 1763), 8.

II.8 The third sound

The hypothesis of coexistent sounds lost much of its attractiveness after the discovery of the phenomenon of the third sound: two strong continuous musical tones sounding together may produce an impression of a third sound, usually lower than either of the original ones. It was described in the 1740s–1750s by such musicians and composers as Georg Andreas Sorge (1703–1778), Jean-Baptiste Romieu (1723–1766), and Giuseppe Tartini (1692–1770), and was usually associated with the name of Tartini. The grave harmonic contradicted the hypothesis of coexistent sounds since the aliquot part of a string responsible for it would be longer than the whole string. William Bewley (1726–1783), a surgeon from Norfolk, offered several alternative explanations for the third sound.[68] In one of them, he followed Mairan's idea that all additional sounds are due to additional waves produced by particles of air. He assumed that when two different particles of air are affected by waves of different frequency they transmit part of their motion to a third particle, which can vibrate with the frequency lower than the original ones. In his other hypothesis, when two tones, for instance, *C* and *e*, sound together they excite in the inner ear not only the fibers tuned in unison to them, but also a longer fiber, the aliquot parts of which resonate to *C* and *e*. The vibration of this longer fiber as a whole will produce the third sound an octave *below C*. Finally, Bewley compared the production of the third sound to that of a mixture of two colors. He was probably unaware that this analogy had already been objected to by Mairan. Mairan's chief argument was that color depends on refrangibility of light and thus is an absolute concept, while a musical interval depends on the choice of the fundamental note and thus is merely a relative notion: for instance, 'the *Re* of the *Opera* could be the *Ut* of the *Chapelle*.'[69] If, he said, *Fa*, *La*, and *Sol* correspond, respectively, to the blue, yellow, and green, then a mixture of *Fa* and *La* should give *Sol*, which contradicts reality.[70]

William Jones of Nayland (1726–1800), a clergyman and scholar, discussed a possibility to place a source of the third sound into the ear: "Whether there is a third sphere of undulations, produced in the air by the concurrence of the other two, is a question of some difficulty; but without supposing it, the effect must be referred rather to the imagination than the senses."[71] As being opposed to "senses," "imagination" implies something, which has no connection with reality. This idea did not appeal to scientists, and they preferred Lagrange's hypothe-

[68] [W. Bewley], "*Principles and Power of Harmony* by B. Stillingsfleet" *Monthly Review* 45 (1771): 371–2.

[69] Mairan, 56.

[70] Ibid., 53–4.

[71] [W. Bewley], "W. Jones's *Physiological Disquisitions*" *Monthly Review* 66 (1782): 13.

sis that the third sound is a specific case of beats: when the difference of frequencies of two sounds is small one perceives their mixture as beats but when it is sufficiently great the impression is of a new tone. Although unclear mechanically, Lagrange's theory correctly predicted the frequency of the grave harmonics.[72]

II.9 Summary

How close did the eighteenth-century acousticians come to the principle of superposition of waves and acoustical interference? First, they knew the phenomena connected with these concepts (standing waves in a string or a tube, the beats and grave harmonics). Secondly, some of them adopted the idea of sharing motions. Thirdly, Bernoulli applied the graphical method to add vibrations. Thus, all the components of Young's discovery were known long before him. However, no one attempted to combine them together. Scientists were content with the principle of superposition of motions and felt no need in specifying it for waves. Although Bernoulli's diagram (Fig. 9) reminds one an addition of vibrations in Young's diagrams (Fig. 4), it could not have led him to the principle of superposition of vibrations. One reason was that he added only very specific functions (harmonics), which made it impossible to obtain a periodical pattern and thus suggest a possibility of interference. Another reason was that, unlike Young, Bernoulli was not interested in studying a variation of the amplitude of compound vibration in time; his real purpose was to approximate a continuous function by a trigonometric series.

There were several reasons for hindering the discovery of superposition of waves and interference of sound. First, acousticians were primarily preoccupied with independence of crossing sound waves and not with their interaction. Most of them explained independence of sound without using superposition of motions, and those few (Lagrange and Euler) who did apply this concept did not achieve much. Secondly, the contemporary theory of hearing supported the view that independence of sound waves is incompatible with their interaction. Thirdly, theoretical acoustics was then primarily a theory of music and thus was concerned with frequencies and not with amplitudes. The pulse model of sound correctly predicted the frequency of a compound vibration but was useless for studying its amplitude. Fourthly, the problem of interaction of waves was oversimplified: it was reduced to an interaction of two oscillators which have the same or the opposite phase. In many cases the problem was solved without any

[72] Lagrange, 142–44.

actual addition of vibrations at all. For instance, physicists concluded that two vibrations of commensurable frequencies reinforce one another and those of incommensurable frequencies extinguish each other, solely from the consideration that there was a resonance in the former case and no resonance in the latter case. Finally, there was no adequate mathematical means to deal with the addition of sound. Mathematicians abandoned Newton's harmonic representation of sound as too particular. Two alternatives were left: the pulse model, favored by physicists, and very general functions introduced by D'Alembert and Euler. None of them fit the problem: the former was too simple, while the latter was too complicated.

Young succeeded in his discovery because he adopted a new approach to the problem of the intersection of waves. Previously, scientists were concerned with the explanation of the fact that two waves remained the same *after* their intersection as they were *before* it. Young concentrated instead on the *process of interaction* itself. Thus, when stating that intersecting waves change one another he, in fact, did not contradict his predecessors: they were simply talking about different stages of the same process.

The difference between the two approaches probably appeared to Young too subtle to claim a major discovery. Indeed, his terms sounded like those of Euler and Lagrange, and he did not explain any new phenomenon that had not been explained before. Apparently, he claimed his investigation of the variation of the amplitude of compound vibrations in time to be an innovation. This includes the principle of superposition of vibrations and the principle of superposition of waves for waves of different frequency. He also discovered some conditions of coherence for such waves. Thus, we may attribute to Young the discovery of the principle of temporal interference of sound.

It is important to indicate here an interdependence between the concepts of superposition and interference. It appears from our definitions in Ch. I that the discovery of the principle of superposition of waves must preceed that of the principle of interference. However, from the practical point of view, at the time when no one was concerned with amplitudes of vibrations, one could not have begun to study changes of the amplitude without a good reason. This means that the hypothesis of interference had to appear first, and then it triggered the process that led to the principle of superposition of waves and then to the principle of interference. More exactly, all these ideas had to emerge at the same time, because Young needed to go from one to the others back and forth more than once until they all became clear to him, and until he obtained theoretical results conformable with observations.

Let us now see whether Young applied the principle of superposition of waves and the principle of interference to intersecting waves of the *same* frequency.

Chapter III

Young and interference of mechanical waves

One may suppose that the most natural and easy transition from temporal interference of sound to spatial interference of light (discovered in May 1801) would have been through spatial interference of sound or water waves. The reasoning behind this suggestion is that it is easier to formulate the principle of spatial interference and the conditions of coherence for mechanical waves, and then to transfer them without significant changes to light. It will be shown, however, that facts defy this logic. I will also discuss in this chapter the response to Young's mechanical concept or interference.

III.1 Standing waves

The simplest case of spatial interference is a standing wave, which is a superposition of the direct and reflected waves. Apparently, Young knew this, to some extent, in 1799. In his "Outlines" he says that a node in a string is formed by "successive impulses, reflected from the fixed points at each end, destroying each other."[1] To "destroy" one another the "impulses" must be continuous functions; this is also confirmed by Young's diagram of the wave produced by a bow applied near an aliquot point, which reminds a standing wave.[2]

Apparently, the phenomenon of the "resonant passage" provides another example of a standing wave:

> When the walls of a passage, or of an unfurnished room, are smooth and perfectly parallel, any explosion, or a stamping with the foot, communicates an impression to the air, which is reflected from one wall to the other, and from the second again towards the ear, nearly in the *same direction* with the primitive impulse: this takes place as frequently in a second, as double the breadth of the passage is contained in 1130 feet [1130 feet/sec is the velocity of sound]; and the ear receives a perception of a musical sound, thus determined in its pitch by the breadth of the passage ... If the sound is predetermined, and the frequency of vibrations such as that each pulse, when doubly reflected, may coincide with

[1] Young, "Outlines," 90.
[2] Ibid., 87, Fig. 111.

the subsequent pulse proceeding directly form the sounding body, the intensity of the sound will be much increased by the reflection.[3]

In the first part of this passage, Young refers to a then popular view that a periodical sequence of discontinuous short noises makes a musical tone.[4] It is essential that, unlike Sauveur's "pulses," the Young's ones posses a direction. In the second part, however, he deals with a continuous sound, which the observer hears reinforced wherever he stands, which could happen if the direct and reflected waves are such as in Fig. 10.[5] There is no indication that Young ever experimented with spatial interference of sound after 1799.

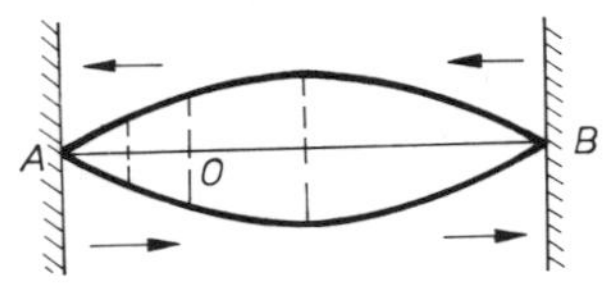

Fig. 10
An illustration to Young's "resonant passage"

His first known experiments on spatial interference of water waves were performed as lecture demonstrations at the Royal Institution in April 1802, using a glass-bottomed ripple tank.[6] He described this apparatus and some of the experiments on interference and diffraction of water waves in his *Lectures* (Fig. 11).[7]

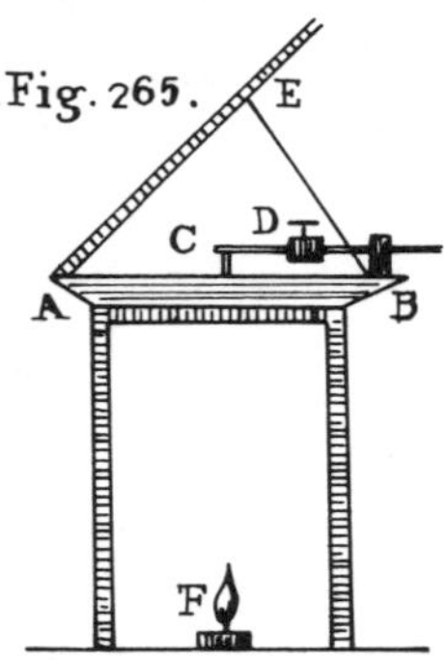

Fig. 11
Young's ripple tank apparatus (from *Lectures* 1: Fig. 265)

[3] Ibid., 73. Italics added.
[4] See, for instance, Lagrange, 145.
[5] If the observer stands at *O*, the sound reflected at *A* and *B* travels the distance $OA + AB + BO = 2AB$ before it reaches the ear. Thus, only the sound of the frequency $f = v/2AB$ (v is the velocity of sound) will resonate, in accordance with Young.
[6] Young, *Journals of the Royal Institution* 1 (1802): 201–202.
[7] Young, *Lectures* 1: 290, 777, plate XX, Figs. 265–267.

Here he gave a clear account of a standing wave formed by a wave striking against a steep wall (Fig. 12):

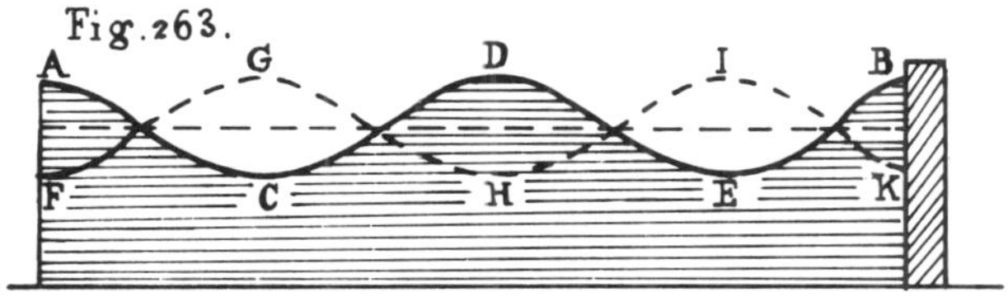

Fig. 12
A standing wave on the surface of water produced by reflection from a wall (from Young's *Lectures* 1: Fig. 263)

> When this reflection is sufficiently regular, *it is easy to show*, that the combination of the direct with the reflected motions must constitute a vibration, of such a nature, that the whole surface is divided into portions, which appear to vibrate alternately upwards and downwards, without any progressive motion, while the points which separate the portions remain always in their natural level.[8]

The caption to the figure adds that "elevation and depression become however twice as great as before reflection."[9] The words "it is easy to show" imply that Young derived an equation for the standing wave.

Thus, Young's own experiments on spatial interference of water waves which could have led him to optical interference, took place *after* its discovery. Let us now see whether he could rely on the work of others, in particular, of Newton.

III.2 Tides

It has been frequently suggested that Newton discovered interference of water waves.[10] This opinion is based on Newton's explanation of irregular tides at the Chinese port Batsha:

[8] Ibid., 290, Figs. 263, italics added.
[9] Ibid., 777.
[10] See the prehistory of Newton's account in I. B. Cohen, "First explanation of interference," *American Journal of Physics*, 8 (1940): 104.

> Further, it may happen that the tide may be propagated from the ocean through different channels towards the same port, and may pass quicker through some channels than through others; in which case the same tide, divided into two or more succeeding one another, may compound new motions of different kinds. Let us suppose two equal tides flowing towards the same port from different places, the one preceding the other by six hours; and suppose the first tide to happen at the third hour of the approach of the moon to the meridian of the port. If the moon at the time of the approach to the meridian was in the equator, every six hours alternately there would arise equal floods, which, meeting with as many equal ebbs, would so balance each other, that for that day, the water would stagnate and be quiet.[11]

Young first referred to this passage in the syllabus of his lectures on natural philosophy, written after his "Theory od light," which does not metion any tides at all.[12] He said about the role of Newton's explanation in the discovery of interference of light:

> there was nothing that *could have* led to it by any author with whom I am acquainted, except some *imperfect hints* in . . . the works of the great. Dr. Robert Hooke, which had never occured to me at the time that I discovered the law; and except the Newtonian explanation of the combinations of tides in the Port of Batsha."[13]

Thus, Young is saying that Newton's explanation "*could have* led" to his discovery but not that it "led" him. Although he first read *Principia* in 1790, it is not necessary that Young recalled Newton's account of the tides at Batsha when working on interference of light. In all probability, it caught his eye after the discovery was made: in the fall of 1801, he was engaged in an extensive reading to prepare his course, and he could have reread *Principia.*

But even if Young were aware of Newton's account, he could hardly have learned much from it. According to our definitions, the passage from *Principia* does not contain the principle of interference of waves. Indeed, spatial interfer-

---

[11] Newton, *The Mathematical Principles of Natural Philosophy*, trans. by A. Motte, rev. by F. Cajory (Univ. of California Press: Berkeley, 1946), 439.

[12] Young, *A Syllabus of a Course of Lectures on Natural and Experimental Philosophy* (London, 1802), 143. The book was printed in January 1802. Young was appointed the Professor of Natural Philosophy of the Royal Institution in July 1801. At the time he was composing his "Theory of light," which was ready no later than August 1801 (see n. 2, Ch. VI). This means that he was working on the *Syllabus* in the fall 1801, after finishing the "Theory of light."

[13] Young, "Reply," 203, italics added.

ence is recognizable by a permanent pattern of several alternating crests and troughs. Since a tide has a very large wavelength (thousands miles), only an extraterrestrial observer can see this interference pattern. On the other hand, the passage does not contain the principle of superposition of waves either, for it mentions only one minimum and says nothing about other maxima and minima. Since the two motions are added only at one point, the concept involved is the principle of superposition of motions.

This interpretation agrees with Newton's description of tides in general:

> The two luminaries excite *two motions*, which will not appear distinctly, but between them will arise one *mixed motion* compounded out of both. In the conjuction or opposition of the luminaries their *forces* will be conjoined, and bring on the greatest flood and ebb. In the quadratures the sun will raise the waters which the moon depresses, and depress the waters which the moon raises, and from the difference of their *forces* the smallest of all tides will follow.[14]

Young interprets this in his *Syllabus* as a case of temporal interference: "The solar tide is about two fifths of the lunar tide, and the joint effect of both is to produce a periodical increase and diminution, in the same manner as the undulations of sound are combined with each other."[15]

It is more likely, in my view, that Newton's 'motion' is a displacement of a water particle from its position of equilibrium under the action of either the sun or the moon. Since he added the two motions only at particular positions of the sun and the moon, the only principle he needed was the principle of superposition of motions. In fact, the motions are vibratory, and one may argue that because Newton found the conditions at which the compound vibration has its smallest and greatest amplitude he anticipated the principle of superposition of vibrations. But even if he did, there is no reason to attribute to Newton the principle of superposition of waves.

First, even assuming that in both passages cited above Newton meant waves, he considered a *single* system of waves and their interaction only at a *single* point of space. Secondly, Newton never claimed to discover any *new* principle different from that of the superposition of motions. Had he realized that it was a new idea, he could have applied it to other wave phenomena, which he never did. Nor did anyone else between Newton and Young. Laplace, for instance, based his

[14] Newton, 436, italics added.
[15] Young, *Syllabus*, 143.

theory of tides on the solution of the wave equation, in which the role of the sun and the moon was accounted for without any superposition of waves.[16]

Young was the first who interpreted tides in terms of interference. It was easy for him to do this *after* he discovered the interference of light; however, if tides were his only guide, his chances for the discovery would have been negligible. Indeed, since the two tidal waves could not be isolated, it was impossible to *prove* directly that a tide was a compound motion. Nor could this be done indirectly by proving the truth of Newton's theory of tides, for this theory had not been considered then as fully demonstrated.

Thus, apparently, it was not spatial interference of water waves realized in the standing wave or in tides which led Young to the principle of interference of light. I will now show that the same is true for spatial interference of sound.

III.3 Coherence of mechanical waves

To formulate the principle of interference for any case of spatial interference Young had to discover the relevant conditions of coherence. One cannot do this by studying the standing wave, where these conditions are satisfied automatically. Thus, even if Young had known by 1799 more of standing waves than he revealed in his "Outlines," that would not have been of much help. Let us see whether he tried to derive the conditions of coherence from other phenomena of interference of sound.

There was one condition, which Young considered to be applicable to any phenomena of acoustical interference: the condition of direction. He said in 1802 that in his objections to the coalescence of two sounds Gough "thinks proper to omit the *only* case in which I have asserted its existence that is, when the sounds arrive at the ear 'in the same direction.'"[17] Initially, he understood by "direction" the one from the ear to a sounding body. Soon he found, however, that beats could be heard even when the directions to two sounding bodies were very different, as, for instance, when one body was a tuning-fork held between the teeth and the other was outside the ear. Thus, Young had to modify the condition: sounds were to coincide in direction only *inside* the "ultimate organ of hearing."[18] Perhaps, that was how he explained Tartini's observation that the grave harmonics could be produced even when the two instruments were quite distant from one another.[19]

[16] Laplace, *Mécanique céleste* 5 vols (Paris, 1800–1825), 2: 222–27.

[17] Young, "An answer to Mr. Gough's 'Essay';" 265.

[18] Young, "Observations in reply to Mr. Gough's Letter on the grave harmonics," *Nich. Jour.* 4 (1803): 74.

[19] Young referred to Tartini in his "Outlines." Even if he did not read Tartini's book, he could find this experiment in M. Young's book (see *Enquiry*, 185), which he certainly read.

In his *Lectures*, Young gave the following explanation of the condition of direction:

> When the two sounds, thus propagated together, coincide very nearly in direction, the motions belonging to each sound may be resolved into two parts, the one in the common or intermediate direction, and the other transverse to it; the latter portions will obviously be very small; they will sometimes destroy each other, and may always be neglected in determining the effect of the combination, since the ear is incapable of distinguishing a difference in the directions of sounds which amounts to a very few degrees only.[20]

This may be understood in two ways. One is that an increase of the angle between the two waves reduces the longitudinal component of the compound velocity, which in turn makes the maxima less noticeable. The other one is that the sounds coming from different directions will be perceived separately, and this may prevent interference. The latter argument implies again that interference takes place inside the ear. Young never demonstrated experimentally that coherent sounds must have almost the same direction. His theoretical explanation of this condition was not very convincing either. It is not improbable that he merely extended to sound the condition he had known to be true for light.[21]

Another source of confusion for Young was the condition of a common origin. As shown in I.4, two waves are coherent if their coherence time is comparable with the length of observation and much greater than the resolution time of the detector. These limitations can be overcome for sound and water waves, and one can see or hear the interference pattern produced by two independent sources. It appears from his scarce references to interference of sound of equal frequency (spatial interference) that Young was unaware of these restrictions.

For temporal interference of sound two sources easily produce coherent waves, and Young was correct in his explanation of intersecting sounds of different frequency (the beats or grave harmonics). Unfortunately, he decided that interference will remain when the difference of frequencies approaches zero. To demonstrate that sounds of the same frequency but of the opposite phase destroy one another he referred to the effect made by two strings tuned almost in unison: after beats became very faint the sound suddenly disappeared.[22] This explanation is false: the sound was lost not because of interference but because it passed the limit of audibility.

---

[20] Young, *Lectures*, 1:389, italics added

[21] It will be shown further that Young knew most conditions of coherence for light as early as 1801.

[22] Young, "Observations," 73.

Somehow Young missed that his result contradicted that of Smith, who also experimented with two sounds of the same frequency and opposite phase.[23] According to Smith, such sounds were produced by two halves of the same string, and they made a perfect harmony. If Young applied his principle of superposition of waves to Smith's experiment, his conclusion would have been correct, for the two halves of the string are coherent sources, and thus destroy sound at some points of space. This effect was, however, masked by resonance in the wooden parts of the violin, which produced sounds of the same frequency but with different phase relations.[24] Young, of course, did not know that; and having no other satisfactory explanation of Smith's experiment, he might have decided to ignore it.

After 1799, Young did not do any experiments on spatial interference of sound. His statements on the subject probably came from extrapolating the results obtained for temporal interference. Actually, some of these statements describe the principle of superposition of waves rather than the principle of interference of sound.[25]

Interestingly, Young held in his hands an excellent device for demonstrating spatial interference of sound, but he used it for an entirely different purpose. In 1801, he discovered that the intensity of sound of a tuning fork depends on the direction of observation.[26] He did not recognize, however, that the effect was due to interference of sound, produced by two prongs. This was discovered by Wilhelm Weber (1804–1891) only in 1826.[27]

Young was more successful with water waves. He correctly described in his *Lectures* the interference of water waves produced by two stones of the same size simultaneously dropped into a pond.[28] He also gave a precise drawing (Fig. 13) of a spatial interference pattern created by two vibrators and pointed out that lines of maxima and minima are hyperbolae.[29] This remark proves that the diagram was based on calculations rather than on observations. Perhaps, such was also

---

[23] See Ch. II, n. 25.

[24] I owe this explanation to Professor M. Blair, of the University of Minnesota.

[25] Young, *Syllabus*, 92.

[26] Young, "Bakerian lecture. On the theory of light and colours," *Misc. Works* 1 : 150. First publishes in *Phil. Trans.* 1802 : 12–48. Henceforth is referred to as "Theory of light."

[27] Weber, "Ueber Unterbrechungen der Schallstrahlen," *Journal für Chemie und Physik* 48 (1826): 385–430.

[28] Young, *Lectures* 1 : 290. These two wave systems are sufficiently coherent to create an interference pattern lasting several seconds.

[29] Ibid., 1 : 777, Fig. 267.

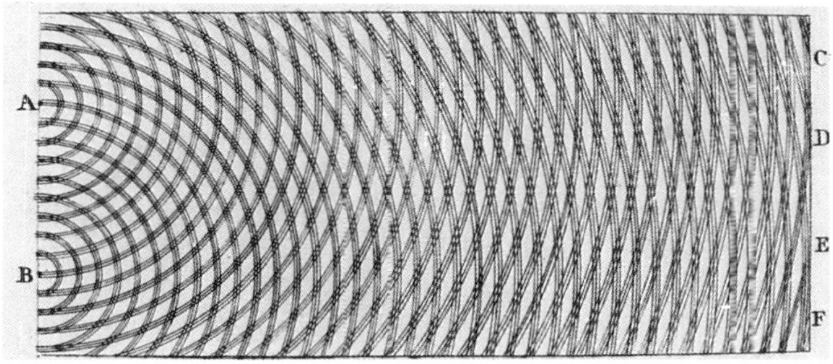

Fig. 13
Spatial interference of waves (from Young's *Lectures* 1: Fig. 267)

the origin of his other note that the centers of interfering waves must be close to one another, which is an equivalent of the condition of direction.[30] There is no evidence that Young systematically studied interference of water waves for its own sake, thus, he probably derived the shape of the loci of maxima and the conditions of coherence by analogy with interference of light.

It appears that after 1799 Young lost any interest in studying interference of mechanical waves and fully concentrated on optics. However, the first challenge to his concept of interference came not from optics but from acoustics.

III.4 The response to interference of sound

a) Robison

Young's article "Outlines" was received quite favorably both in Britain and abroad.[31] This does not mean however an approval of the concept of coalescence of sound, which was either ignored or criticized. Robison, for instance, did not mention it at all even when commenting on Young's criticism of Smith:

---

[30] Ibid., 290.

[31] The article was reprinted in *Nich. Jour.* 5 (1802): 72–78, 81–91, 121–28. An abridged translation into the French appeared in *Bibliothèque Britannique* 14 (1800): 210–26, 301–29. A complete translation into the German by Vieth was published in *Ann. der Phys.* 22 (1806): 249–85, 337–92. It was reviewed in *Monthly Review* 33 (1800): 260–63; and in *Annals of Philosophy* 1 (1801): 41–43.

> We are therefore surprised to see this work of Dr. Smith greatly undervalued, by a most ingenious gentleman in the Philosophical Transactions for 1800, and called a large and obscure volume, which leaves the matter just as it was, and its results useless and impracticable. We are sorry to see this; because we have great expectations from the future labours of this gentleman in the field of harmonics, and his late work is rich in refined and valuable matter.[32]

Here Robison defends primarily Smith's method of tuning musical instruments by means of beats. In other places, however, he touches some theoretical points too.

Robison repeats Smith's arguments against Sauveur that consonances have nothing to do with coincidence of vibrations. In particular, he says that "*coalescence* of sound, which makes the pleasing harmony of a fifth, for example, does not arise from the coincidence of vibrations."[33] Young's term "coalescence" may suggest that the passage was directed against Young. Among Robison's proofs one is of a special interest to us. He says that experience shows that two organ pipes set, say, ten feet apart and sounding the same note C produce a perfect harmony in any part of the room. However, according to the theory of coincidences, such a harmony should take place only at specific points, such as the midpoint between the pipes. If an observer, Robison continues, will approach one foot nearer to one of the pipes, then every pulse of one pipe will bisect the pulse from the other. Since such pulses never coincide, the observer must not perceive any harmony, which contradicts observations.[34]

Interestingly, this explanation is totally inconsistent with Robison's explanation of the beats. He calls them "an alternate enforcement and *diminution* of the strength of sound"[35] Bearing in mind that Young was the first who drew attention to the periodical *weakening* of sound during the beats, this passage suggests Young's influence. This impression becomes even stronger from Robison's account of beats, in which he says that if two pulses coincide, "the agitation pro-

---

[32] Robison, "Temperament of the scale of music," in the *Supplement to the Encyclopaedia Britannica*, 3rd ed. (Edinburgh, 1801); references are given to the reprint in Robison's *System of Mechanical Philosophy*, 4:411–2.

[33] Ibid., 406, italics added.

[34] Ibid. According to Robison, the frequency of the tone C is 240 Hz, which means that its wavelength is 4 to 4.5 feet, depending on the chosen velocity of sound. When the observer moves 1 foot off the center the path difference of two waves will be 2 feet or one half of the wavelength. Thus, according to Young's theory such waves would destroy one another, and the observer would hear nothing at that point. Although he did not say it directly, Robison might have considered this argument as refuting Young.

[35] Ibid., 408, italics added.

duced by one pulse is increased by that produced by the other," while if the pulses bisect each other, "the agitations of the one will *counteract* or *weaken* those of the other."[36] This sounds much closer to Young than to any of his predecessors. Nonetheless, while adopting Young's explanation of the beats of sound, Robison does not want to accept his concept of interference and its role in music. Young replied to Robison that "the misstatement, relative to the non-interference of different sounds, is an inaccuracy which far outweighs the merit of Dr. Smith's share of this improvement [of tuning]."[37] He ignored Robison's theoretical excursions altogether, some as irrelevant and others (the experiment with two pipes) because he was unable to answer them. However, when attacked on a similar ground by Gough, he decided to reply.

b) Gough

John Gough (1757–1825), a blind natural philosopher of Kendal and a tutor of John Dalton, was a scientist of reputation, although not of the stature of Robison. Like Robison, he rose to defend Smith against Young.[38] In Gough's view, Young's concept of coalescence contradicts both mechanics of sound and its perception. First, he attacks the idea of sharing motions.[39] He considers an air particle, affected by two waves traveling in different directions. The compound displacement of the particle is a diagonal in a parallelogram whose sides are the displacements produced by either wave separately. If the waves have different frequency, the ratio of the sides of the parallelogram will change with time and so will the direction of its diagonal, which, to Gough, is the direction of the transmitted wave. Since, in his view, one perceives sound wave as having a particular direction only if it retains this direction for some time, Young's concept of "coalescence" is wrong.

Young replies that Gough's argument is irrelevant, because it ignores that coalescing sounds must have almost the same direction.[40] In his turn, Young also attempts to show that his rival's concept has no mechanical foundation:

> If the mind were capable of making up a sound in the way that Mr. Gough supposed, we ought to hear, whenever the impulses of one

[36] Ibid., 408–9, italics added.
[37] Young, "A letter . . . respecting sound and light," 135.
[38] Gough, "The theory of compound sounds," *Memoirs of the Literary and Philosphical Society of Manchester* 5, pt. 2 (1802): 653–65; also *Nich. Jour.* 4 (1803): 152–9.
[39] Ibid., 155–7.
[40] See n. 17.

> sound bisect either accurately or very nearly, the intervals between the impulses of another sound, an imaginary note, an octave above the separate sounds: if, on the contrary, my opinion is true, we must conclude that the retrograde motions of the one will counteract the direct motions of the other, and that both the sounds will be destroyed.[41]

The following diagram illustrates Young's idea (Fig. 14). Let sound be represented by a wire curve which moves uniformly to the right so that its peaks strike

Fig. 14
An illustration of Young's interpretation of Gough's argument: the frequency of the compund sound is the number of times both sound waves (represented by the moving wires 1 and 2) strike the ear-drum E every second.

an elastic plate *E*. Let us further assume that the ear "counts" the number of vibrations in the same way as the plate *E* does to strokes. Then, in the case of a "mixture" of two sounds of about the same frequency the number of strokes produced by both curves *AB* and *CD* will be double that made by either one, which means that the compound sound must be perceived as an octave to the primary ones. Since this never occurs in reality, Young concludes, a "mixture" of sounds does not correspond to a system of independent waves.

Young suggests the following realization of his thought experiment:

> we have two sounds standing in this relation, in the intervals between the beats of two musical chords tuned very nearly in unison. And if we listen to a grave sound which beats very slowly, while it is dying away, we shall observe, that in the interval of the last and faintest beats, when the sound is least mixed by reflections and irregular propagations, the note, instead of rising to the octave, is wholly lost.[42]

In fact, Young's argument is not very convincing, since decay of sound can explain the phenomenon as well as interference. Moreover, he does not know why the same effect is not observed when both strings are tuned *exactly* in unison.

[41] Young, "Observations," 73.
[42] Ibid.

To Young, the third sound is the best demonstration of interference of sound, but Gough points out several flaws in it. His first point is that the intensity of the third sound is *always* less than the intensities of primary sounds, while if Young's theory were true, the amplitude of a compound sound would have been alternately greater and weaker than those of the primary sounds.[43] To this Young replies that, generally, to acquire the same direction, two sound waves must undergo several reflections, which considerably reduces their intensity.[44] (It was due to his exchange with Gough that Young decided that vibrations must coincide inside the ear as mentioned above.) Another point is that the third sound has no definite direction and, as Young put it, "rings in his ears."[45] Gough retorted that if this is the case, the third sound is not a real sound but only an "effort of the imagination."[46]

This does not mean that Gough objects to any involvement of hearing in the explanation of acoustical phenomena. He simply does it differently than Young. For instance, Gough understands coalescence as mingling of several sounds in a single new entity. To him, a compound sound is only a mixture in which different sounds can be easily detected, while "were it possible for a number of sounds to coalesce, and form but one, the compound would acquire sensible properties *perculiar to itself*, and at the same time lose the distinguishing characters of its elements."[47] Gough compares the perception of compound sounds with that of physico-chemical compounds: in a "mixture" of sounds the components preserve their nature as vinegar and pepper do in the infusion of pepper in vinegar; while in a "coalescence" of sounds the properties of the primary sounds are lost as the taste of soda or of the muriatic acid is lost in common salt. Thus, since the ear can distinguish individual tones in a compound sound, the concept of coalescence is false.

In his reply, Young indicates that although in some cases of coalescence the ear perceives a new property (a new tone), it also recognizes the original sounds. To resolve the paradox he supposes that,

> The ear indeed appears to have greater powers of analysis than one would naturally have expected, it decomposes a "compound" just as if it were a mere "mixture", not only in this case [of the third sound] but in

43 Gough, "On the nature of the grave harmonic," *Nich. Jour.* 4 (1803): 3.
44 Young, "An answer to Mr. Gough's 'Essay'," 266.
45 Young, "Letter from Thomas Young . . . in reply to Mr. Gough's Letter, at p. 36 of the present volume. On the phenomena of sound," *Nich. Jour.* 3 (1802): 146.
46 See n. 43.
47 Gough, "The theory of compound sounds," 153–4, italics added.

> *many others*: how it performs this operation, I do not pretend to determine.[48]

Perhaps, he conceived this idea while thinking of the mathematical expression of simple and compound sounds. As early as 1799, Young adopted Bernoulli's hypothesis that every periodical function can be representend by a sum of harmonics and even tried to determine the amplitudes of harmonics for the triangular function.[49] In 1801, he noted in his *Syllabus* that, to represent sound, a sinusoid "seems indeed to have some natural claim to preference, for the ear generally *analyses* more complicated vibrations into such subordinate ones as may be derived from this form."[50] It was more than a speculation, for Young found that "there is an *experimental* reason to believe, that the *purest and most homogeneous* sounds do in fact agree very nearly with the law of this [sine] curve."[51] Possibly, he referred to his experiments on recording vibrations: a small pencil fixed at the end of a vibrating rod marked a sinusoidal line on a sheet of paper drawn along.[52]

Young does not explain the "other" phenomena, and I will speculate about his possible reasoning in two cases. First, if each of two vibrating strings produces a simple vibration, two waves enter the ear and create a compound vibration in the eardrum. The latter communicates it to the inner ear, which being aided by the brain "decipheres" it and produces the sensation of two pure tones: the same tones, which would have been perceived if each string sounded separately. Thus, an intersection of waves does not affect the sound. Second, let a vibrating string make a compound vibration. This vibration is transmitted to the ear and copied by the eardrum. In the inner ear the compound vibration is resolved into several harmonics. Thus, harmonics do not exist outside the ear, they are produced inside it.

Young considers the third sound to be an objective phenomenon, that is that an air particle, affected by two primary sounds, acquires a compound vibration which is communicated to the ear, and the latter "decomposes" it producing the third sound. He presumes that the ear is an "ordinary" (that is, linear) oscillator. If that were true, his analogy between the beats of sound and the third sound would be correct. Fifty years later, however, Helmholtz indicated that the analogy is false because the ear is a non-linear oscillator and thus a sum of two primary vibrations produces a number of vibrations of different frequencies, per-

[48] See n. 44, italics added.
[49] Young, "Outlines," 85.
[50] Young, *Syllabus*, 91, italics added.
[51] Young, "An account of Dr. Young's harmonic sliders [1802]," *Misc. Works* 1:216, italics added.
[52] Young, *Lectures* 1:191.

ceived as "combination tones."[53] Since non-linearity was unknown to Young's contemporaries, they could have agreed to his analogy provided they accepted his hypothesis of the analytical power of the ear. Gough, however, was not willing to do that. In his view, "this power cannot be admitted *before* the doctrine of coalescence is established."[54] With little evidence to support his hypothesis, Young did not press it further. It was forgotten until Georg Ohm (1787–1854) rediscovered it forty years later.[55]

At the end of the controversy, Gough accepted the idea of sharing motions, although he never agreed that that was the cause of beats of sound and of the third sound. In general, the controversy ended inconclusively. In one respect, it was natural, since with physiology of hearing in a rudimentary stage neither side could prove anything involving perception of sound. On the other hand, the disputants could have avoided a number of erroneous statements if they relied on actual experiments rather than on the thought ones. Young, for instance, could have proved that the low intensity of the third sound had nothing to do with multiple reflections of sound.

c) Later response

The controversy between Young and Gough attracted attention to the concept of interference. In 1803, in his review of Gough's "Theory of Compound Sound" Brougham calls Young's concept of coalescence a "mathematical absurdity."[56] On the other hand, he also dislikes Gough's theory because of its confusion of the physical and psychological aspects of sound. However, Brougham's own theory of the perception of a mixture of sounds is even more confusing that those of Young and Gough.

In 1805, Gerhard Ulrich Anton Vieth (1763–1836), a professor of mathematics from Dessau, published his comment of the Young-Gough controversy, supplemented by excerpts from their articles.[57] In his view, the main point of the debate is the theory of the third sound. Vieth thoroughly analyzes all evidence presented by Young and Gough and concludes that the phenomenon of the third sound is true but its interpretations by both sides are deficient. He argues that Gough's calling the third sound a "product of imagination" is wrong, for *every* sound is a

---

[53] Helmholtz, "Ueber Combinationstone," *Ann. der Phys.* 99 (1856): 497–540.

[54] Gough, "Reply to Dr. Young's letter on the theory of compound sounds," *Nich. Jour.* 3 (1802): 40, italics added.

[55] Ohm, "Ueber die Definition des Tones . . .," *Ann. der Phys.* 59 (1843): 496–565.

[56] [Brougham], "Memoirs of the Philosophical Society of Manchester," *Edingburgh Review* 2 (1803): 192–6.

[57] Vieth, "Ueber Combinationstone, in Beziehung auf einige Streitschriften über die zweier englischer Physiker, Th. Young und Jo. Gough," *Ann. der Phys.* 21 (1805): 265–314.

product of subjective perception and as such can be called "imaginary." He agrees with Young that the third sound results from compound vibrations of the air particles, but denies any resemblence between this phenomenon and those of tides or harmonic sliders. To Vieth, a combination of two sine curves does not represent the third sound, since in case of a the fifth the compound curve does not resemble a sinusoide and thus cannot represent a pure tone. He strongly objects that unisons, produced by two strings, can destroy one another if the pulses of one bisect the intervals between the pulses of the other.[58]

Other acoustical authors, such as Jean-Henri Hassenfratz (1755–1827), Chladni, René-Just Haüy (1743–1822), Simon-Denis Poisson (1781–1840), and Jean-Baptiste Biot (1774–1862) did not take any notice of interference at all.[59] The subject of interference of sound was brought back in 1816 in connection with Fresnel's principle of interference of light but its understanding did not improve. Like Gough, Biot stated that the third sound could be explained without any interference, since it was only a "sensation."[60] Poisson recalled a case when a simultaneous firing of many guns caused silence: in his view, it was an example of destructive interference.[61] The first systematic study of interference of sound was undertaken by Wilhelm Weber in 1826.[62]

III.5 Summary

Thus, neither the principle of superposition of waves nor the principle of interference were known before Young. The main reason for that was that scientists had not been looking into right direction: they have been preoccupied with independence of waves and paid little attention to the change of amplitude. In those few cases when they did it (Newton, for instance), they did no go beyond adding two motions at selected points or at specific moments of time. While others were concerned with comparing waves *before* and *after* their intersection, Young

[58] Ibid., 306, 312–13.

[59] Hassenfratz, "Sur la cause que augment intensité du son dans portevoix," *Ann. de Chim.* 50 (1805): 297–311; "Mémoire sur la propagation du son," *Ann. de Chim.* 53 (1805): 64–74; Chladni, *Die Akustik* [1802] (Leipzig, 1830); Poisson, "Mémoire sur la théorie du son," *Journal de l'École Polytechnique* 7 (1808): 319–92; Biot, *Traité de physique expérimentale et mathématique* 4 vols (Paris, 1816), 2: 114–16.

[60] "Mélanges. Notice des Séances de l'Acad. R. des Scienc. de Paris," *Bibliothèque Universelle* 1 (1816): 305–6.

[61] Ibid., 306.

[62] See n. 27.

inquired what occured *during* the intersection. He found that under special conditions the intersecting waves produced a regular effect which could be noticed.

Young did not realize the importance of a formal distinction between different principles of superposition. Having been confused by a similarity between his terms and those of his predecessors ("concurrence of vibrations," "breaking of waves," and others) he exaggerated their contribution. This could account for Young's incertitude about the novelty of his ideas. Although he did not have these specific terms Young distinguished between superposition of waves and their interference, at least, in the case of temporal interference of sound and spatial interference of water waves. An early switch to optics prevented Young from discovering spatial interference of sound.

Young was ill-prepared for the debate on interference of sound. He did not know that the conditions of coherence were not the same for sounds of identical and different frequencies. Apparently, after his discoverey of interference of light he felt no compulsion to investigate further interference of sound. It appears that he tried to extrapolate the principle of interference of light to spatial interference of sound and water waves. This approach backfired when he entered a controversy with Gough. Young wanted to use interference of sound to illustrate interference of light but he failed this task. Hastily selected arguments, some erroneous and other unverifiable and all unsupported by accurate experiments, did not convert Gough. The choice of the third sound as the crucial argument in favor of the concept of interference was the most unfortunate, because the confusion between physical and physiological aspects of this phenomenon could not be resolved at the time. Besides, Young was unaware that it was a nonlinear effect and thus could not be compared to interference of water waves or to interference of light. Had he concentrated on spatial interference of water waves instead, he would have prepared a much better ground for optical interference. As it was, the dispute rather weakened Young's quest for optical interference (see VIII.2)

Young left the interference of sound and of water waves undeveloped (and so it remained until 1825–26, when physicists began studying them by analogy with interference of light) because he discovered optical interference. Let us now see what he had to do to make this discovery.

Chapter IV

Discovery of the principle of interference of light

As stated in I.3, Young wanted to build a mathematical theory of periodical colors. Before examining this theory, let us see how this problem had been considered before Young.

IV.1 Optical background

The foundation of physical optics was laid down in the second half of the seventeenth century by Robert Hooke, Francesco Grimaldi (1618–1663), Christiaan Huygens, and Isaac Newton. Huygens' influence on Young is quite evident. Young considered him to be one of the founders of the wave theory of light, for he spoke of it as Huygens' theory.[1] This theory was based on the so called "Huygens' principle."[2] According to this principle, to construct the wave front at any given moment it is sufficient to take the wave front at some preceding moment, and consider it as a locus of innumerable centers of spherical secondary waves; the common tangent surface to all these secondary waves is the new wavefront. Huygens' theory of double refraction in Iceland spar, based on his principle, at first received several favorable comments but shortly afterwards was almost completely forgotten.[3] This probably happened because his theory ignored the phenomena of colors and did not give any mechanical explanation of Huygens' principle. Young knew Huygens' principle and accepted his account of the rectilinear propagation of light. However, he underestimated the role of this principle in the explanation of other optical phenomena. He believed, for instance, that Huygens' demonstration of the laws of reflection and refraction of light was no better than those of Barrow or Euler.[4]

Young acknowledged Hooke not only as one of the founders of the wave theory of light but also as the author of a very ingenious hypothesis on the colors of

---

[1] Young, "Outlines," 79.

[2] Huygens, *Treatise on Light*, 19–21.

[3] Alan E. Shapiro, "Kinematic optics: a study of the wave theory of light in the seventeenth century," *Archive for History of Exact Sciences* 11 (1973): 245–46.

[4] Young, "Theory of light", 151–52, 155. See also Shapiro's comparison of Huygens' explanation of refraction with those of his predecessors in his "Kinematic optics", 207–8.

thin films.[5] Hooke considered light to be a periodical sequence of pulses. When pulses such as *ab* (Fig. 15) fall on a plane-parallel plate, some of them *(cd)* are reflected from the first surface *AB*, while the other pulses *ef*, after two refractions and one reflection at *EF* emerge parallel to *cd*. Evidently, the pulses *cd* are stronger than *ef*. According to Hooke, when the stronger pulse precedes the weaker one, the two having a distance between them less than that between two subsequent incident pulses, such a compound pulse produces a sensation of yellow. When the weaker pulse precedes the stronger one, a blue color is observed, and so on.[6]

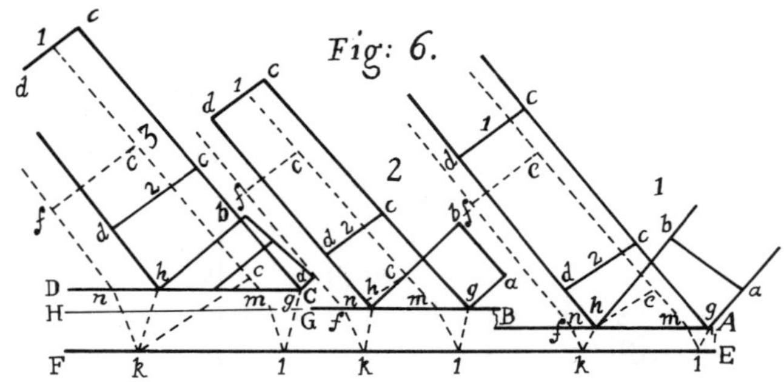

Fig. 15
Hooke's explanation of the colors of thin films, which involves two waves reflected from two surfaces of the film (from Hooke, *Micrographia*)

To some physicists in the early nineteenth century Hooke's explanation, and especially his diagram, appeared very similar to Young's. In fact, Hooke's account had nothing to do with interference, and he never discussed the superposition of motions. His "duplicated" pulse was a complex of two independent waves rather than a single wave formed by the superposition of two original ones. On the other hand, Young admitted that if he had known it before his discovery of the principle of interference, it "might have led me earlier to a similar conclusion."[7]

Though Hooke demonstrated that the color depends on the thickness of a thin plate, he was unable to find a quantitative relation between them. This problem

[5] Young, "Theory of light", 141, 162.
[6] Hooke, *Micrographia* (London, 1665), 65–6. Hooke knew somehow that a ray of light which experienced more refractions or reflections loses a greater part of its original intensity.
[7] Young, "Theory of light", 162.

was solved by Newton, who found experimentally that the thicknesses of a thin plate in the places where colored rings of successive orders are seen form an arithmetic progression. To interpret this empirical law, Newton assumed that a light ray when entering a refracting medium, undergoes periodic alternations of its state which predispose the ray to be reflected or transmitted at the second surface.[8] He called these alternations "fits of easy transmission and easy reflection." On the basis of his measurements in thin films, Newton ascribed to light a specific constant of space periodicity (the "intervals of fits").[9] This interval, he found, depended on color, on the index of refraction of the medium in which the light propagated, and also on the angle of incidence of the light entering this medium. Using the concept of the interval of fits Newton developed a mathematical theory, which when applied to the phenomenon of the colors of thick plates showed a good agreement with observations.[10] We should note that in Newton's theory the concept of periodicity alone was insufficient for calculating the positions of the colored fringes. Newton therefore had to introduce a supplementary hypothesis: whatever the initial direction of the incident light, it experiences the same number of fits between two parallel surfaces. In other words,

$$\frac{e_r}{I_r} = \frac{e}{I} = m, \tag{4.1}$$

where e_r and e are the routes passed by the oblique and normal rays, respectively, while I_r and I are the intervals of fits in these two directions (see my Fig. 16), m is the number of fits. When light experiences an even number of fits in traversing the plate, it will pass through the plate, but if an odd number, then the light will

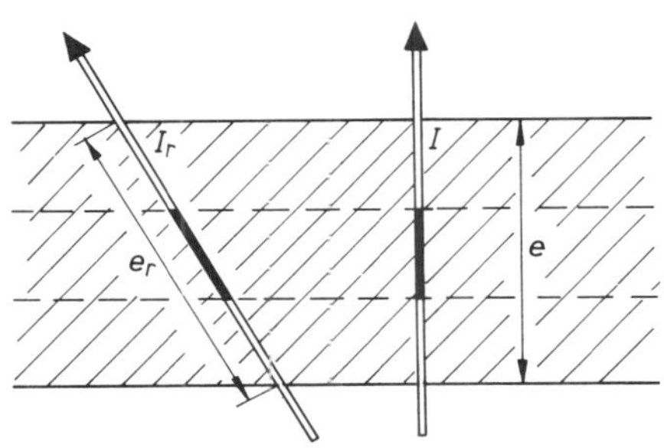

Fig. 16
An illustration of Newton's theory of fits of easy reflection and easy transmission

[8] Newton, *Opticks* (New York, 1952), 200–203. First published in 1703, this facsimile is of the 1730 edition.
[9] Ibid., 278–81.
[10] Ibid., 281–88, 292–93, 299.

be reflected back. Therefore, to determine what will happen to the light at the plate, given its thickness and the angle of incidence, one needs to know both the length of the ray's path, e_r, and the interval of its fits I_r (in the case of a plane-parallel plate one may take e and I, which are constants). This supplementary hypothesis was readily applicable to bodies with parallel or almost parallel surfaces. It was more difficult to use it for spherical drops of water in clouds, which, according to Newton, were responsible for the halos around the sun and the moon.[11] Perhaps Newton considered only those rays that passed almost in a radial direction, and in this case the drop's diameter will be the "length *e*." It seems that Newton also tried to find an equivalent for this length for diffraction, but he failed and left this phenomenon without a quantitative explanation.

Newton's great achievement was to prove that in several optical phenomena light revealed the same constant of periodicity. This fact was the core of his theory of fits, and later it became the basis for Young's principle of interference. Young preferred Hooke's periodic light waves to Huygens' aperiodic waves as the only model reconcilable with the principle of interference. He found that the wavelengths of these waves were double of Newton's intervals of fits at normal incidence, which provided him with the constants necessary to develop a quantitative theory. Euler's ideas of periodical light waves also influenced Young. In his theory of the colors of thin films Euler assumed that the difference of colors of light was due to the difference in the frequency of vibrations of the ether.[12] To explain the periodic alternation of colors seen on the surface of a thin plate he supposed that, like vibrating strings, the particles of bodies had their specific (resonant) frequencies of vibration depending on their size. When set in vibration by light falling on a film, these particles emit light whose frequency depends on the particle's size, and different places of the film send to the eye light of different colors (apparently, Euler identified the thickness of the thin film with the size of "particles").[13]

Newton claimed that his theory of fits was empirically based and did not need any mechanical interpretation. For those who liked mechanical models he offered the following hypothesis of the fits' origin: when a light ray strikes the surface of a body, it excites waves in the ether that pervades all bodies. These waves travel faster than light, and by condensing and dilating the ether at different distances from the surface, they create at the given distance conditions favorable for the subsequent reflection or transmission of light.[14] This hypothesis did not ap-

---

[11] Ibid., 313–14.

[12] Euler, "Essai d'une explication physique des couleurs engendrées sur des surfaces extrement minces," *Hist. Acad. Berlin* 8 (1752): 262–82.

[13] Ibid., 277.

[14] Newton, *Opticks*, 280.

peal to scientists. Those few who did accept Newton's concept of fits, preferred to give it different mechanical interpretations.[15] The majority of physicists, however, denied this concept as mechanically inconsistent. Whether Newton believed in his explanation of fits or not, he at least applied it as a mathematical hypothesis, but no one followed him. Though Newton's observations of thin films were praised as an example of his experimental skill, his mathematical theory of fits was completely abandoned. This occurred because eighteenth-century opticians felt no need to study the colors of thin and thick plates quantitatively. They were concerned only with such questions as whether these colors are produced by one of the two refracting surfaces (and by which one), or by an unknown "agent." [16]

A similar approach prevailed in the sudy of diffraction. Both Grimaldi and Hooke studied diffraction qualitatively, being concerned with a demonstration that it was a new property of light.[17] Newton attempted to find a mathematical relation between the dimensions of the colored fringes and the distances of the diffracting body from the source of light and observation screen; unfortunately, he was interrupted and did not complete this work.[18] Giacomo Filipo Maraldi (1665–1729) and Joseph-Nicolas Deslisle (1688–1768) also made similar measurements, but they never tried to derive any mathematical law from their results.[19] The most popular qualitative explanations of diffraction in the eighteenth century were those invented by Newton which agreed with the emission hypothesis of light. According to one of them, diffraction (or "inflection" as Newton called it) was a kind of refraction of light which took place in an ethereal atmosphere surrounding all bodies.[20] Mairan proposed a similar hypothesis in

---

[15] Joseph Priestley discussed several models of fits in his *The History and Present State of Discoveries Relating to Vision, Light, and Colours* (London, 1772), 308.

[16] Mazeas, "Observations sur des couleurs engendrées par le frottment des surfaces planes et transparent," *Hist. Acad. Berlin* 8 (1752): 248–61; Dutour, "Recherches sur le phénomène des anneaux colorés," *Mem. Sav. Étrang.* 4 (1763): 285–312; and Duc de Chaulnes, "Observations sur quelques expériences de la quatrième partie du deuxième livre de l'Optique de M. Newton," *Hist. Acad. Paris*, 1755: 136–44.

[17] Grimaldi, *Physico-mathesis de lumine, coloribus et iride* (Bononiae, 1665), 1. Hooke presented his papers on diffraction to the Royal Society in 1672 and in March 1674/5; see Birch, *History of the Royal Society*, 4 vols. (London, 1756–57), 3: 10, 194–95. Hooke's clearest account of diffraction is in *The Posthumous Works of Robert Hooke* (London, 1705), 187–90.

[18] Newton, *Opticks*, 317–38. See also Roger H. Stuewer, "A critical analysis of Newton's work on diffraction," *Isis* 61 (1970): 188–205.

[19] Maraldi, "Diverses expériences d'optique," *Hist. Acad. Paris*, 1723: 157–200; Delisle, "Expériences sur la lumière et les couleurs," in *Mémoires pour servir à l'histoire et au progres de l'astronomie, de la geographie et de physique* (St. Petersbourg, 1738), 205–66.

[20] Newton's hypothesis was read before the Royal Society on December 16, 1675. See Birch, *History fo Royal Society* 3: 268–9; reprinted in *Isaac's Newton's Papers and Letters on Natural Philosophy*, ed. I. Bernard Cohen (Cambridge, Mass; 1958), 198–9.

1738, probably independently of Newton.[21] Later, Étienne-François Dutour (1711–1784) modified Mairan's hypothesis, and, in particular, added the possibility of the reflection of light at the boundaries of the ethereal atmosphere and the body's surface.[22] In Newton's other hypothesis the bending of rays near a body resulted from an interaction between light and the body. He assumed that the force of interaction increased as the distance between the ray and the body decreased.[23] This hypothesis, with various modifications, was widespread in the eighteenth century. Both hypotheses were vague enough to account for a number of phenomena of diffraction of light, yet they were useless for quantitative purposes, such as for some astronomical observations.

All of these models of diffraction provided only qualitative explanations of the phenomena, but the "undulationists" did not have even a qualitative one. When Grimaldi brought forth an analogy between diffraction of light and a flow of water around an obstacle, scientists objected to it as contradicting the existence of shadows. It must be noted that the "emissionists" (Newton) considered diffraction as produced by an external agent (an ethereal atmosphere or a force), while "undulationists" (Grimaldi and Hooke) believed that diffraction was due to a natural tendency of light revealed when light rays passed very near a body. Young later explored both approaches.

The situation with the study of double refraction in the eighteenth century was similar to that of periodic colors. In their attempts to simplify the account of this phenomenon as much as possible, physicists abandoned Huygens' theory of double refraction.[24] One might suppose that Huygens' theory was discarded due to its complexity and association with the unpopular wave hypothesis of light. However, Newton's very simple theory, published in his widely read *Opticks*, was forgotten as well.[25] René-Just Haüy in 1788 was the first who compared both theories, he concluded that Huygens' theory better agreed with observation.[26] How-

---

[21] Mairan, "Recherches physico-mathématiques sur la réflexion des corps. Troisième partie. De la diffraction," *Hist. Acad. Paris*, 1738: 53–64.

[22] Dutour, "De la diffraction de la lumière," *Mem. Sav. Étrang.* 5 (1768): 635–77.

[23] Newton, *Opticks*, 339.

[23a] On how the study of diffraction affected the dispute on the nature of light see N. Kipnis, "History of the study of diffraction of light in the 18th century" (in Russian) in A. Bogolyubov (ed.), *Mekhanika i fizika XVIII v* (Moskow: Nauka, 1976), 213–27.

[24] De la Hire, "Sur les réfractions d'une espèce de talc," *Hist. Acad. Paris*, 1710: 341–52; Leclerc Buffon, *Histoire naturelle génerale et particuliere*, 125 vols. in 61 (Paris, 1799–1808), 13: 272–83. Giambatista Beccaria, "Observations sur la double réfraction du crystal de Roche," *Journal de Physique* 2 (1772): 504–10.

[25] Newton, *Opticks*, 354–8.

[26] Haüy, "Sur la double réfraction dans la spath d'Island," *Hist. Acad. Paris*, 1788: 34–60. The qualitative character of the eighteenth-century study of double refraction may be seen from Jed. Z. Buchwald's critical analysis "Experimental investigation of double refraction from Huygens to Malus", *Archive for History of Exact Sciences* 21 (1980): 311–73.

ever, Haüy's article was also ignored. When in 1802 William Hyde Wollaston (1766–1828) measured the index of refraction of the extraordinary ray in Iceland spar, he was unaware of the existence of a mathematical theory of double refraction.[27] It was just a coincidence that he had a friend such as Young, who was able to inform him about Huygens' theory.

The qualitative approach to physical optics was rather typical of the physical sciences in general, especially in the first half of the eighteenth century. However, in the second half of this century a trend appeared towards the mathematization of physics. Mechanics became a model for the development of acoustics, electrostatics, and other branches of physics.[28] Nonetheless, the level of mathematization of these sciences by the end of the century still remained relatively low. Most experimenters, though concerned with improving the precision of measurements, had no interest in deriving mathematical laws from their observations. On the other hand, mathematicians usually were indifferent to experimental verification of their physical theories. Very few scientists combined experimental skill with an interest in mathematical study of phenomena. In optics several areas were mathematized, such as photometry, atmospheric refraction, and the theory of aberration of fixed stars. However, such fields as the theory of periodical colors and double refraction remained mostly qualitative.[29] A brief discussion of the work of Brougham, Gibbs Walker Jordan (1757–1823), and Wollaston (all of whom Young mentioned in his "Theory of light") will give us an idea of the methodology used in physical optics in Young's time.

In 1796–97 Brougham published the results of his research on diffraction and double refraction performed during his undergraduate studies at the University of Edingburgh.[30] He reported several interesting observations, but in this chap-

[27] Wollaston, "On the oblique refraction of Iceland crystal," *Phil. Trans.*, 1802: 381–7.

[28] See, for instance, John Heilbron on the quantification of electrostatics in his *Electricity in the 17th and 18th Centuries* (Berkeley, 1979), 444–89; and Frankel, "J. B. Biot and the mathematization of experimental physics in Napoleonic France," in *Historical Studies in the Physical Sciences* 8 (1977): 34–41.

[29] David Rittenhouse's study of the diffraction grating was an exception. He correctly derived from observations that the angular shift of the colored fringes from the central white fringe is proportional to the order of the fringes (see "An optical problem, proposed by Mr. Hopkinson, and solved by Mr. Rittenhouse," *Transactions of the American Philosophical Society* 2 (1786): 201–6. Young made no references to Rittenhouse, probably because he had not read his paper. William Nicholson gave a short account of some of Rittenhouse's observations and added a few experiments of his own, together with a qualitative explanation of them ("A remarkable effect of the inflection of light passing through wire cloth, not yet clearly explained," *Nicholson's Journal* 1 (1797): 13–6). However, Nicholson did not mention Rittenhouse's quantitative result, and thus Young could not learn about it from this paper.

[30] Brougham, "Experiments and observations on the inflexion, reflexion and colours of light," *Phil. Trans.*, 1796: 227–77; and "Further experiments and observations on the affections and properties of light," ibid., 1797: 352–85. See also Steffens' discussion of these papers in his *Newtonian Optics*, 86–92.

ter I will focus only on some of his theoretical views. Brougham followed Newton in an attempt to mathematize the study of diffraction.[31] He attributed to light a new inherent property, "different flexibility," (an analog of Newton's "different refrangibility"), which was supposed to explain the different bending of light rays of different colors passing near a body. He distinguished three types of flexibility: inflexibility, deflexibility, and reflexibility. The first two described diffraction in transmitted light (rays bend toward the body or away from it, respectively), while the latter dealt with diffraction in reflected light. Brougham deduced from his observations that "the flexibilities of the rays are inversely as their refrangibilities."[32] While the difference between inflexibility and deflexibility was vague, his concept of different reflexibility was expressed quite clearly: "The angle of reflexion is not equal to that of incidence, except in particular (though common) combination of circumstances, and in the mean rays of the spectrum."[33] In other words, he stated here that the law of reflection is not a universal one, but rather a specific case of diffraction. Pierre Prevost misunderstood this result and strongly criticized Brougham for an apparent violation of a fundamental law of optics.[34] Brougham believed that all phenomena of diffraction could be accounted for by a simple equation analogous to Snell's law:

$$\frac{\sin i}{\sin r} = \text{const}, \qquad (4.2)$$

where the constant depends on the color of light, i is the angle of incidence, and r is the angle of diffraction. Brougham's evidence was too limited to justify this mathematical law (he presented two experiments of his own and one of Newton); nonetheless, his attempt to establish quantitative relations between diverse phenomena of colors deserves attention. He was the first who discovered an analogy between the colors of diffraction in reflected light and those of thin films.[35] His attempt to link the concepts of different reflexibility and different refrangibility was also significant. However, Brougham could not mathematize his concept of different reflexibility, for he had rejected Newton's fits.

Jordan also aimed at improving Newton's results on diffraction.[36] Although he repeated all of Newton's observations, systematically changing the size of the

[31] Brougham, "Experiments and observations," 227.
[32] Ibid., 277.
[33] Ibid.
[34] Prevost, "Quelques remarques d'optique, principalement relatives à la réflexibilité des rayons de la lumière," *Phil. Trans.* 1798: 311–31. See also Paul A. Tunbridge, "An unpublished paper on light by Lord Henry Brougham F.R.S. (1778–1868)," *Archives des Sciendes* 26 (1973): 111–17.
[35] Brougham, "Further experiments," 269–71.
[36] Jordan, *The Observations . . . Concerning the Inflection of Light* (London, 1799).

diffractor and its distances from both the aperture (the source of light) and the screen, he did not take any measurements. Jordan pointed out, in opposition to Newton, that diffraction fringes could be observed not only outside the shadow of a body ("external fringes"), but also inside it ("internal fringes"). This fact was well known to Grimaldi and Maraldi, but it was forgotten afterwards. Jordan also believed that he proved that diffraction was due to an attraction of light toward bodies and not a repulsion, as Newton had thought. Young utilized this result in his first account of diffraction.

Wollaston was closer in his method to Brougham than to Jordan: he made measurements and compared them with a theory when he had one. He had no particular interest in physical optics and came to the study of double refraction by a chance, after inventing a new method to measure refraction and dispersion of light by total reflection.[37] While applying this method to various substances Wollaston came across Iceland spar. He discussed with Young the results of his measurements of refraction of the extraordinary ray and learned of Huygens' theory from him. Then he compared his observations with theoretical predictions and found that they "correspond more nearly than could well happen to a false theory."[38] Thus, several of Young's contemporaries were involved in physical optics, and some of them attempted to quantify it. Let us now see whether they influenced the formation of Young's scientific method.

Although Young is famous for his extensive knowledge of scientific literature, before his discovery of the optical principle of interference he did not read much on physical matters. His first acquaintance with optics took place before 1786 when he read Benjamin Martin's *Lectures on Natural Philosophy*. He applied his knowledge of geometrical optics to build microscopes and telescopes.[39] In 1790 he thoroughly studied Newton's *Principia* and *Opticks*, but some passages in the latter left him unsatisfied.[40] Young returned to optics while studying the physiology of vision. He published a paper on this subject, which brought him membership in the Royal Society of London.[41] Between 1796 and 1799 he read extensively on acoustics and related mathematics and was particularly impressed by the works of Continental mathematicians. It is unlikely that Young had read much on physical optics before 1799, for in his "Outlines of experiments" he mentioned only Newton, Euler, Jordan, and Huygens.[42] Young's reference to

---

37 Wollaston, "A method of examining the refractive and dispersive powers, by prismatic reflections," *Phil. Trans.*, 1802: 265–80.

38 Wollaston, "Oblique refraction," 382.

39 Peacock, *Thomas Young*, 5, 17.

40 Young, "Reply," 197.

41 Young, "Observations on vision [1793]," *Misc. Works* 1: 1–11.

42 Young also mentioned in his "Observations on vision," Smith and Petrus van Musschenbroek, but these authors had nothing on periodic colors of interest to Young.

"Huygens' theory of light" does not prove that he already was familiar with the *Traité*, but two years later, in the "Theory of light," his references to Huygens became much more specific. I believe that he read Huygens, as well as many other optical authors, in the summer and fall of 1801, when he prepared his course of lectures on natural philosophy at the Royal Institution. Thus, Newton's *Opticks* determined the direction of Young's research in physical optics; Euler inspired his interest in the wave hypothesis of light; and Newton, Euler, Bernoulli, Lagrange, Smith, and other mathematicians were responsible for his mathematical approach to physics. By the time of his discovery of the principle of interference Young had already shown himself to be proficient in both experiment and mathematics.[43]

There is a misconception among historians of science that Young's dislike of the higher branches of mathematics prevented him from developing a more sophisticated theory of light than he actually did. Young, in fact, read through the most difficult mathematical treatises of his predecessors and contemporaries, yet he believed that every physical problem had to be resolved in the simplest mathematical and experimental way. If he considered geometric methods to be sufficient, he did not use differential equations, and so on.[44] Perhaps, on some occasions a more general mathematical method would have yielded more general results, but his mathematics in the case of the principle of interference seems to be fully adequate. In that case he had to *simplify* his mathematical means rather than complicate them (see section 3 of this chapter).

IV.2 The transition from acoustics to optics

Young presented his theory of periodical colors in his article "Theory of light," read to the Royal Society on November 12 and 19, 1801. Proposition VIII of this article says:

> When two undulations, from different origins, coincide either perfectly or very nearly in direction, their joint effect is a combination of the motions belonging to each.[45]

---

[43] See Young's papers written between 1799 and 1801: "On the mechanism of the eye," *Misc. Works* 1: 12–63; "An essay on cycloidal curves," ibid., 314–20; and "Lettre . . . sur le formule pour refractions," *Bibliothèque Britannique* 18 (August 1801): 354–63.

[44] Young first used fluxions to solve an optical problem in his "Mechanism of the eye" in 1800. He mentioned an application of fluxions to a geometrical problem in his "Observations on vision."

[45] Young, "Theory of light," 157.

Historians have claimed that this proposition expresses the essence of Young's discovery in optics and called it "the principle of interference of light."[46]

In fact, Young does not assert that Proposition VIII is limited to light. To the contrary, he extends it to sound: "I have, on a former occasion, insisted at large on the application of *this* principle to harmonics, and it will appear to be of still more extensive utility in explaining the phenomena of colours."[47] The tendency to cover different kinds of waves is common to several Propositions in the "Theory of light," because Young wants to prove that light has the same properties as other waves.[48]

The real problem, though, is not so much with the name of the discovery as with its content. Most historians do not specify the content at all.[49] Others believe that Young began to think about the interference of waves of light *after* he presented his "Outlines of experiments."[50] I intend to defend an alternative view, namely, that Young was thinking of interference of light waves as early as 1798–99, and that his 1801 discovery was the next stage of the development of the optical principle of interference. At that time he arrived at the law for the path difference of two interfering rays of light and some of the conditions of coherence.

Young first revealed his interest in the nature of light in his "Outlines". In section X, "On the analogy between light and sound," he compares the emission and wave systems of light.[51] Though he does not claim a definite position in this controversy, his selection of arguments shows him to be a proponent of the wave system. He supports it by referring to the analogy between light and sound, and he discusses this analogy in other sections of the paper as well. For instance, he shows that sometimes sound also diverges very little.[52] He also supposes the same law of decay for both sound and light and questions the validity of Lagrange's law that the intensity of sound is inversely proportional to the distance from a source.[53] When comparing the arguments Young set forth in 1799 and in 1801 in favor of the wave hypothesis, we can see changes only in those on the nature of colors. But, if in 1799 Young was already convinced of the existence of

---

46 Whewell, *History*, 3: 93; Steffens, 121; Worrall, 138; Morse, "Young, Thomas," in *Dictionary of Scientific Biography*, ed. by C. Gillispie, vol. 14 (New York: Scribner & Sons, 1975), 566; E. Frankel, "Jean-Baptist Biot: the career of a physicist in ninethenth century France," (Ph.D. diss., Princeton University, 1972), 296.

47 Young, "Theory of light," 157, italics added.

48 Ibid., 148, 149, 151, 154, 156.

49 See, for instance, Whittaker, 101; and Silliman, 95.

50 Steffens, 114, 117; and Latchford, 132.

51 Young, "Outlines," 78–83.

52 Young, "Outlines", 73–75.

53 Ibid., 76.

waves, why did he not try then to apply his hypothesis of interference to light? In fact, it appears that he did so.

There is an obscure passage in the "Outlines" which has been neglected by historians:

> The strength of the joint sound is double that of the simple sound only at the middle of the beat, but not throughout its duration; and it may be inferred, that the strength of sound in a concert will not be in exact proportion to the number of instruments composing it. Could any method be devised for ascertaining this by experiment, it would assist in the comparison of sound with light.[54]

At that time Young assumed that the intensity (the "strength") of sound is proportional to the velocity of vibrating particles. Thus, to find the intensity of a compound vibration produced by two sounds of equal strength and almost the same frequency, he simply adds the graphs representing the two sounds. It is clear that the resulting curve will be of double amplitude only where the phase of both sounds is the same, and less than that in other places (see Young's Fig. 93 in Fig. 4). Hence, the strength of the compound sound would be less than double the strength of each of its components. Now, it was generally accepted in the eighteenth-century acoustics that *m* identical instruments placed near one another would produce a sound *m* times as strong as any one of them (a similar assumption was used in photometry). The passage just cited implies a loss of a part of the total energy caused by the interference of two sounds. Young realized that his statement contradicted the traditional view and suggested that it be checked experimentally. In his opinion, such an experiment "would assist in the comparison of sound with light." Since all the other analogies between light and sound in his paper are based on their *similar* properties, I believe this holds true for this case too. It implies that by 1799 Young admitted that waves of light could destroy one another. Additional evidence for this conclusion comes from Young's later recollections of his Cambridge period (1797–99):

> In making some experiments on the production of sounds, I was so forcibly impressed with the resemblance of the phenomena that I saw, to those of the colours of thin plates, with which I was *already acquainted*, that I began to suspect the existence of a closer analogy between them than I could before have easily believed.[55]

54 Ibid., 84.
55 Young, "Reply", 199, italics added.

Young is apparently referring to the analogy developed in the "Outlines" in which he suggested that the phenomenon of the colors of thin films where "the same color recurs whenever the thickness answers to the terms of an arithmetical progression" was "precisely similar to the production of the same sound, by means of an uniform blast, from organ-pipes which are different multiples of the same length." [56] In Young's opinion, this analogy justifies Euler's idea that color depends on the frequency of ether vibrations. Young possibly borrowed this analogy from Euler but interpreted it differently. Euler compares color to a pitch and believes that as organ-pipes of different length can produce the same note, so thin plates of different thicknesses can form the same color. In Young's view, there is one-to-one correspondence between color and frequency of vibration of the ether (or the interval of fits in the vacuum), and plates whose thicknesses contain an odd number of intervals of fits reflect the same color.

In other words, Young is implying the existence of a standing wave of light. One can see this in the following passage:

> Supposing white light to be a continued impulse or stream of luminous ether, it may be conceived to act on the plates as a blast of air does on the organ-pipes, and to produce vibrations regulated in frequency by the the length of the lines which are terminated by the two refracting surfaces. It may be objected that, to complete the analogy, there should be tubes to answer to the organ-pipes: but the tube of an organ-pipe is only necessary to prevent the divergence of the impression, and in light there is little or no tendency to diverge; and indeed, in the case of a resonant passage, the air is not prevented from becoming sonorous by the liberty of lateral motion.[57]

Thus, according to Young, two parallel surfaces are necessary for a formation of a standing wave of sound or light, and the frequency of vibrations depends only on the distance between the two surfaces. It is worth noting that a color is formed not by a periodical primary vibration but by a "continued impulse or stream", just as in Bernoulli's theory where a musical tone is produced by a "blast of air." This implies that Young represents light mathematically exactly as he does sound, i. e. by an arbitrary continuous function which can be resolved into harmonics (this is obviously borrowed from Bernoulli). In modern terms, Young considers a thin refracting plate to analyze light in the same way as an organ pipe analyzes sound. In both cases the produced frequency depends only on the dis-

[56] Young, "Outlines", 82.
[57] Ibid.

tance between the two surfaces. In postulating standing wave of light it would have been natural for him to associate its antinodes with bright fringes and nodes with dark fringes. Since he knew that a node is formed where two vibrations destroy one another, this implies that he accepted the possibility of the destruction of light. Young's analogy was in fact another interpretation of Newton's explanation of the colors of thin plates: to determine whether the given plate would reflect or transmit the given light, Newton counts the number of intervals of fits contained between two surfaces, while Young does the same with the half-wavelengths. At the time Young probably already suspected that Newton's interval of fits at normal incidence of light was equal to the half-wavelength, and there is no doubt that Young's explanation was based upon Newton's.

As applied to thin films, Young's optical-acoustical analogy produced no new results compared to Newton's theory of fits. However, Young dared to extend it to diffraction by a narrow body, where Newton failed. According to Young, every body is surrounded by a homogeneous ethereal atmosphere of a finite radius which refracts the rays of light passing near the body.[58] He also assumes that light can be reflected by the surface of the body and by the boundary of the atmosphere. His basic idea is that "the length of the track of a ray of light through the inflecting atmosphere may determine its vibrations."[59] He seems to mean that the ray *ABCD* (see my Fig. 17) will be reflected or refracted at point *C* depending on whether the length *BC* contains an odd or even number of half-wavelengths. Thus, in this model the length of a ray's trajectory inside the ethereal atmosphere

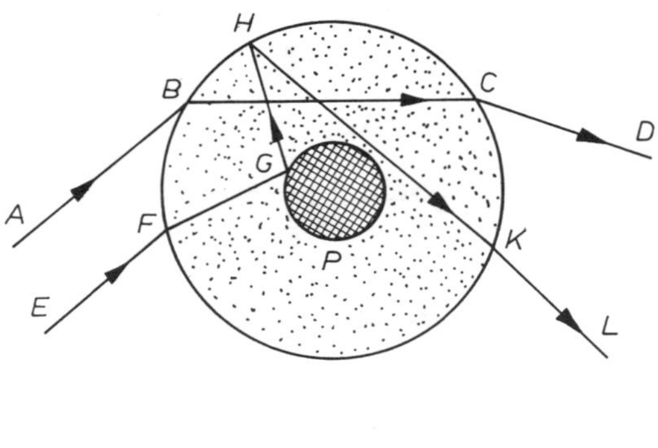

Fig. 17
An illustration of Young's explanation of diffraction from his "Outlines": the length BC or HK of the light ray inside the ethereal atmosphere FBHCK surrounding the body P determines the color as the length of an organ pipe determines the tone.

[58] Ibid.
[59] Ibid.

plays the same role as the "length *e*" in Newton's theory of fits. To account for the periodic repetition of diffraction fringes of the same color Young considers such trajectories as *EFGHKL*, of which the part *FGHK* determines the ray's fate. The major difference between Young's model of diffraction and those of Maïran and Dutour was the condition of transmission or refraction of light which made his model quantitative. There was, however, a notable difficulty for a practical application of the model, since there was no independent method to determine the radius of the ethereal atmosphere or its index of refraction. This possibly explains why Young never tried to obtain numerical results with it. John Herschel and George Peacock believed that Young's optical acoustical analogy was "fanciful." [60] In my opinion, this interesting hypothesis – a synthesis of Newton's and Euler's ideas of the periodicity of light – marked a significant intermediate step toward the wave theory of light. And it was one of the earliest examples of Young's interest in a quantitative study of physical phenomena, particularly in physical optics.

Thus, by 1799 Young considered the possibility of the destruction of light otherwise than by absorption. It was the first step towards the discovery of the optical principle of interference. Young could not explicitly argue for the idea that light added to light can produce darkness. It would have been viewed as contradicting common sense, and Young did not possess evidence sufficient to overcome this belief. He could not have referred to the alternate bright and dark fringes in phenomena of periodical colors as a proof for his hypothesis, for these fringes were then also explained as resulting from a condensation and rarefaction of light rays. As a quantitative physicist, Young had to develop a mathematical theory of interference and compare it to experiment. We do not know what his first step was. Since he started from the temporal interference of sound, I would conjecture that he also began his quantitative study of interference of light with temporal interference (Fresnel likewise discussed temporal interference in his very first paper). If he did, he would have soon realized that temporal interference of light could not occur (see Ch. V). This would explain why in the "Theory of light" he considered interference of waves of *equal* frequency, without mentioning any other possibility. But even if Young was aware of the exclusively spatial character of interference of light as early as 1799, it does not mean that he was already close to the solution of the problem of interference. He had to overcome three major obstacles: 1) finding a model of spatial interference suitable for developing a general mathematical theory, 2) determining a proper mathematical expression for light waves, and 3) discovering the conditions of

[60] Peacock's footnote in Young, *Misc. Works* 1: 81; and letter form John Herschel to Hudson Gurney, in Schweber, *Aspects*, 149.

coherence. Let us now see how Young dealt with the first two problems (the last will be covered in Ch. V).

As we have seen in this chapter, standing waves were the first case of spatial interference explored by Young. This model, however, was not quite appropriate for a quantitative explanation of diffraction. Besides, when applied to thick plates, it encountered the same difficulties as Newton's theory of fits.[61] Thus, Young had to search for other phenomena of spatial interference. He was already familiar with two, namely, the interference of water waves produced by two stones thrown into a pond, and the tides at Batsha, as explained by Newton. In both phenomena two waves, coming from different sources, met at the same point and produced a vibration whose amplitude depended upon the path difference of the waves: maxima or minima of vibrations occured where the path difference was equal to an even or odd number of half-wavelengths, respectively. Either of these phenomena could have provided Young with a clue towards the derivation of the law of path differences, but there is no evidence that he *verified* this law with observations of water waves. Rather, he applied it directly to light. As shown in III.2, the phenomenon of tides at Batsha could not have been very helpful in deriving the concept of spatial interference even if Young remembered it from his first reading of the *Principia* in 1790; he says only that this explanation "*could* have led" him to the discovery.[62] It is possible that he forgot Newton's explanation when he needed it most and then came across it in the fall of 1801 upon rereading the *Principia.* I would speculate that it was rather the very common phenomemon of the interference of water waves produced by two stones that helped Young. However, to find a proper phenomenon was still insufficient: its observation would have been useless until Young discovered an appropriate mathematical expression for light.

IV.3 The mathematical representation of light

Following the leading mathematicians of the eighteenth century D'Alembert, Euler, Bernoulli, and Lagrange, Young believed that for the sake of generality

[61] Any theory of the colors of thin plates which also claimed to embrace the colors of thick plates had to explain why sufficiently thick plates do not exhibit any colors while thin films of the same substance do. See Young's criticism of Newton's theory of the colors of thick plates in Ch. V.

[62] Young, "Reply," 203, italics added.

sound must be expressed by an arbitrary continuous periodic function of time rather than by a sinusoid. This explains the diagrams (Fig. 4) in his "Outlines," in most of which both interfering sounds appear to be expressed by the triangular function.[63] Young probably viewed this function as a compromise between generality and simplicity, sufficient to explain the second grave harmonic. Adding such functions analitically would have been cumbersome, and Young opted for graphical method, which also provided a better opportunity for discovering the interval of periodicity.

A superposition of waves represented by the triangular function could account, in Young's view, for temporal interference of sound. However, this function was useless for spatial interference, and Young's further progress depended on adopting for this purpose the harmonic function.[64] Sinusoidal waves, introduced by Newton, were occasionally applied by physicists, while mathematicians considered them to be an expression only of a very narrow class of waves. Young, in general, sided with mathematicians, although he recognized the convenience of sinusoidal waves to account for beats of sound.

Explicitly, sinusoidal vibrations first appeared in his *Syllabus*:

> In order to examine the effect of a combination of different sounds, we must *assume* some law for the motions of the particles; and none is so *simple* as that of the cycloidal pendulum or of the harmonic curve, which seems indeed to have some *natural* claim to preference.[65]

Several months later he remarked after a discussion of the use of the harmonic law in acoustics that, "It is therefore by far the most *natural* as well as the most *convenient* to be assumed as representing the state of an *undulation in general*."[66] It seems that Young began to realize that for his purpose the description of waves must be simplified, and that the sinusoidal wave was a good idealization.

The hypothesis of harmonic character of light waves was by itself insufficient for solving any practical problem, in addition an assumption had to be made about the dependence of wave amplitude on time. Young believed that the actual

---

63 See my Fig. 5 and n. 12, Ch. II.

64 There are several problems with a superposition of triangular waves. First, a sum of two triangular functions of the same period, generally, is not a triangular function. Second, when adding two vibrations, expressed by such a function, of the same frequency and amplitude but of different phase we obtain a compound function with flat tops, which makes it impossible to establish a proper relation between the path difference and the wavelength.

65 Young, *Syllabus*, 91, italics added.

66 Young, "Harmonic sliders," 216, italics added. See also his "An account of some cases of the production of colours, not hitherto described," *Phil. Trans.* 1802: 387–97; here the references are given to *Misc. Works* 1: 176. Henceforth, this paper will be cited as "Production of colours."

vibrations of sounding bodies could not be of a constant amplitude.[67] It seems that in 1801 he held a similar opinion about the vibrations of the luminiferous ether. In the "Theory of light," after explaining the addition of two undulations of the same or opposite phase, he remarks that, "in intermediate states, the joint undulation will be of intermediate strength; but by what laws this intermediate strength must vary, cannot be determined without further data." [68] Since an addition of two sinusoidal light waves of the same frequency and constant amplitudes is a simple mathematical problem,[69] Young's remark suggests that he was possibly considering damped harmonic waves with different coefficients of damping. However, all the results presented in the "Theory of light" show that he in fact neglected the extinction of waves and treated them as undamped.

Another possibility is that he referred to some non-harmonic functions, such as the triangular function. It is difficult to obtain an analytical expression for a sum of two triangular vibrations of the same frequency and an arbitrary phase difference. On the other hand, by adding these vibrations graphically only when they are either in the same or in the opposite phase, Young could establish the relation between the path difference and the wavelength in the same way as for a sinusoid (see my Fig. 18). There is evidence that even in 1802 he was not certain

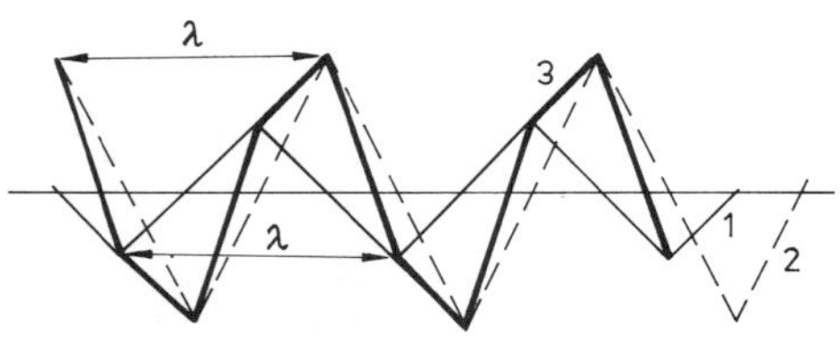

Fig. 18
A superposition of triangular waves of the same frequency

[67] Young, "Outlines," 83.
[68] Young, "Theory of light," 158.
[69] See I.3, eqs. (1.3–1.4).
A damped oscillation may be represented by the equation

$$x = ae^{-\gamma t/2}\cos(\omega t + \alpha)$$

where γ is the coefficient of damping (after time $2/\gamma$ the energy will be e times less than initially). The frequency of damped oscillations ω can be expressed through the frequency of free oscillations ω_0

$$\omega^2 = \omega_0^2 - \frac{\gamma^2}{4}$$

Therefore, if γ is small, the amplitude and the frequency will be about the same as if the waves were undamped.

that the harmonic function is appropriate to represent sound in *all* cases.[70] Thus, it is possible that Young occasionally used such non-harmonic functions for light too, in particular in his "Theory of light." Apparently, he completely abandoned non-harmonic waves only in the *Lectures*.[71]

Hence, if my reconstruction is correct, the sequence of events leading to the discovery of the optical principle of interference was the following. At the time of the discovery of the principle of superposition of waves (1798–99) Young was already thinking of light as waves, while also admitting the possibility of the destruction of light. Possibly he first applied his new principle to temporal interference of light and failed. At that time he knew little about spatial interference, and his view on the mathematical expression of waves could not help him to derive the general law of spatial interference. Consequently, he started from the simplest case, a standing wave, which allowed an arbitrary representation of waves. This model, however, had only a very limited application, and Young turned to the more general case of spatial interference: the interference of two waves of the same frequency coming from two sources in slightly different directions. For such interference the representation of waves by an *arbitrary* periodic function did not work, and Young assumed the waves to be either sinusoidal and effectively undamped, or triangular. In this way be derived the equation for the path difference of interfering waves, which determined the locations of the maxima and minima of vibrations. I suppose that he derived this law when considering the interference of two water waves formed by stones falling simultaneously in a pond.

While summarizing Young's debt to his predecessors, we should note that Newton's theory of the colors of thin films guided him throughout his period of transition from the principle of superposition of waves to the optical principle of interference (at that time Young was not yet aware of Hooke's work). Newton's *Opticks* not only suggested to him the idea of destruction of light, but also provided him with the concept of periodicity of light and with numerical values for the intervals of fits which he identified with half-wavelengths. Newton's account of the tides at Batsha also assisted Young, at least to support his principle. Also, Young may be considered Newton's first successful disciple in the mathematization of periodical colors. Finally, even Young's adherence to the wave hypothesis was partially based on Newton's works. His attempt to support his theory by quotations from Newton has sometimes been considered simply a maneuver to

---

70 Young, "An answer to Mr. Gough's Essay," 266; and "Production of colours," 176 where Young mentions the possibility of a different "law . . . of undulation."

71 Young, *Lectures* 1: Plate XX, Fig. 263; Plate XXV, Fig. 344, 352.

avoid criticism from Newtonians.[72] Sometimes Young has been even accused of trying to present Newton as a supporter of the wave hypothesis.[73] In fact, he never claimed that Newton held to the wave theory, he just insisted on the resemblence between Newton's hypothesis of ether waves and the wave hypothesis of light. He was the first Newton scholar to realize that Newton needed ether waves for his theory of light and colors. Young repeatedly emphasized this idea throughout his life, and this fact is the best proof that his attempt to use Newton's work in the "Theory of light" was not simply a maneuver but an expression of a particular view of Newton's work.[74] Thus, of all of Young's predecessors Newton deserves the most credit for aiding in the discovery of the principle of interference.[75]

IV.4 What is the "law of interference"?

Young called his discovery not the "principle of interference of light" but the "law of interference of light":

> It was in May 1801 that I discovered, by reflecting on the beautiful experiments of Newton, a *law* which appears to me to account for a greater number of interesting phenomena than any other optical principle that has yet been made known.[76]

Since neither term appears in Young's theory of periodical colors in the "Theory of light," and the theory itself is presented in a very sketchy manner, an effort is necessary to separate what refers to all waves from what is specific to light.

It seems that most of what appears in the section "Proposition VIII" is about general properties of waves. For instance:

> When the two series coincide exactly in point of time, it is obvious that the united velocity of the particular motions must be greatest, . . . and

[72] Steffens, *Newtonian Optics*, 120–1; David Hargreave, "Thomas Young's theory of color vision: its roots, development, and acceptance by the British Scientific community," (Ph. D. dis., University of Wisconsin, 1973), 95–7.

[73] Hargreave, "Thomas Young's theory," 96.

[74] Young, *Syllabus*, 96; *Lectures* 1:457, 477; and also "Chromatics," 325.

[75] Steffens, on the contrary, believes that Young "did not use Newton's *Opticks* as his base, and his work was not a continuation of thoughts suggested by Newton" (*Newtonian Optics*, 107). This does not contradict my point, since we are talking about the development of *different* Newtonian traditions.

[76] Young, "Reply," 202, italics added.

> also, that it must be smallest, and if the undulations are of equal strength, totally destroyed, when the time of the greatest direct motion belonging to one undulation coincides with that of the greatest retrograde motion of the other.[77]

According to our definitions, Proposition VIII, together with this note, constitutes the principle of superposition of waves (plus one condition of coherence).

Most instructions on how to apply this principle to light are given in the "Corrolaries" to Proposition VIII, which cover such phenomena as the colors produced by thin and thick plates, by striated surfaces, and by inflection. Let us examine the rules for applying the principle of superposition of waves (for reference purposes, these passages will be marked Y(n) and the date is provided):

Y(1) 1801 The undulations which are now to be compared are those of equal frequency.[78]

> If, therefore, equal undulations of given dimensions be reflected from two points, . . . wherever this line is equal to half the breadth of a whole undulation, the reflection from the depressed point will so interfere with the reflection from the fixed point, that the progressive motion of the one will coincide with the retrograde motion of the other, and they will both be destroyed; but when this line is equal to the whole breadth of an undulation, the effect will be doubled; . . . and if the reflected undulations be of different kinds, they will be variously affected, according to their proportions to the various length of the line which is the difference between the lengths of their two paths, and which may be denominated the interval of retardation.[79]

Thus, we see here the rule for locating maxima and minima and the condition of frequency. (It will be shown further that Young applied then other conditions of coherence as well, although he did not formulate them.) Besides, Young also provides the wavelengths for light of different colors and suggests a phase change for a reflected wave, which travels from a rarer to a denser medium.[80] All this taken together corresponds to our definition of the *principle of interference* for light.

[77] Young, "Theory of light," 157.
[78] Ibid.
[79] Ibid., 158–9.
[80] Ibid., see also V.I.

Thus, originally, Young introduced the principle of interference of light as a consequence of the principle of superposition of waves. However, soon after presenting his "Theory of light" Young began to suspect that chances for a quick recognition of the wave theory were negligible and decided to salvage his new optical principle by offering it as an empirical law. As Young's *Syllabus* shows, this change of tactics occured no later than January 1802.

A specific feature of the *Syllabus* is a separation of facts from theories. The "law of interference" appears among other *facts* in the section "Of Physical Optics," while its explanation is included in the section "Of the Nature of Light." The explanation states: "the general law, by which all these appearances are governed, may be very easily deduced from the interference of two coincident undulations,"[81] while the "law" itself is defined as follows:

Y(2) 1802 Where two portions of the *same light* arrive at the eye by *different routes*, either *exactly or very nearly in the same direction*, the appearance or disappearance of various colours is determined by the greater or less difference in the lengths of the paths: the same colour recurring, when the intervals are multiples of a length, which, in the same medium, is constant, but in different mediums, varies directly as the sine of refraction. In air, this length, for the extreme red rays, is about one 36 thousandth of an inch, and for the extreme violet, about one 60 thousandth.[82]

Here, we have the rule for maxima and minima, at least two conditions of coherence, the relationship between the velocity of light and the index of refraction, and the wavelengths for light of different colors. This is, actually, the same principle as in the "Theory of light," only presented in a more succinct and complete manner. Thus, Young's "law of interference" is, in our terms, the principle of interference of light, and so it will be referred to henceforth.

IV.5 The principle of interference and the theory of interference

Young formulated his principle of interference several more times with some minor changes. Through a comparison of different versions of the principle we can obtain a more clear idea of the structure of Young's theory of interference

81 Young, *Syllabus*, 117–8.
82 Ibid., 113–4, italics added.

and the role of the principle of interference. I will review these versions chronologically.

In his summary of the "Theory of light" for the *Journals of the Royal Institution* the principle of interference is,

Y(3) 1802 a *general law*, that, whenever two portions of the *same pencil* of light arrive at the same point by different routes, the production of colours depends uniformly on the difference of the length of those routes.[83]

In Young's "Production of colours," read before the Royal Society on July 1, 1802, the principle is presented as a quote:

Y(4) 1802 The law is, that "wherever two portions of the same light arrive at the eye by different routes, either exactly or very nearly in the same direction, the light becomes most intense when the difference of the routes is any multiple of a certain length, and least intense in the intermediate state of the interfering portions; and this length is different for light of different colours." [84]

Since the *Syllabus* was the only possible source to quote from, this is a paraphrase rather than a quote of Y(2). (Such way of quoting was not unusual for Young.[85]) In this paper, Young explained the reason for changing his strategy:

> Whatever opinion may be entertained of the theory of light and colours which I have lately had the honour of submitting to the Royal Society, it must at any rate be allowed that it has given birth to the discovery of a simple and general law, capable of explaining a number of the phenomena of coloured light, which, without this law, would remain insulated and unintelligible.[86]

To establish the principle of interference as an empirical law, Young shows its conformity with a number of quantitative and qualitative experiments. The most influential among them became the one, which I call the "screening" experiment: the internal fringes inside the shadow of a narrow body disappear if an opaque screen is inserted into the shadow so as to intercept light coming by one

83 Young, *Journals of the Royal Institution* 1 (1802): 68–9, italics added.
84 Young, "Production of colours," 170.
85 Compare Young's quotation from Smith in his "An answer to Mr. Gough's 'Essay'," 265 with the original Smith, *Harmonics*, 105.
86 Young, "Production of colours," 170.

and only one edge of the body. Young's conclusion was that *two* beams of light are necessary to produce interference.[87]

Young hat nod included any definition of his principle of interference in his "Experiments and calculations," so he made up for this omission when writing an abstract of this paper:

Y(4 a) 1803 — This principle is, that where two portions of light arrive at any point by different routes very nearly in the same direction, they sometimes destroy and sometimes corroborate each other, according to the different lengths of their respective paths.[87a]

New modifications appeared in his *Lectures*:

Y(5) 1807 — In order that the effects of two portions of light may be thus combined it is *necessary* that they be derived from the *same origin*, and that they arrive at the same point by different paths, in directions not much deviating from each other. This deviation may be produced in one or both of the portions by diffraction, by reflection, by refraction, or by any of these effects combined.[88]

The most elaborate formulation was given in Young's article "Chromatics":

Y(6) 1817 — A. – The law is, that when two equal portions of light, in circumstances exactly similar, *have been separated and coincide again*, in nearly the same direction, they will either co-operate, or destroy each other, accordingly as the difference of the times, occupied in their separate paths, is an even or an old multiple of a certain half interval, which is different for the different colours, but constant for the same kind of light.

B. – In the application of this law to different mediums, the velocity must be supposed to be inversely as the refractive density.

C. – In reflections at the surface of a rarer medium, and of some metals, in all very oblique reflections, in diffractions, and in some extraordinary refractions, a half interval appears to be lost.

---

87 Young, "The Bakerian Lecture. Experiments and calculations relative physical optics," *Phil. Trans.* 1804: 1–16. Here the references are given to the *Misc. Works* 1: 179–80. It will be referred to as "Experiments and calculations".

87a Young, *Lectures*, 2: 318.

88 Young, *Lectures* 1: 464, italics added. See also some relevant details in the preceding and following passages.

D. – . . . two portions of light, polarised in transverse directions, do not interfere with each other.

E. – The principal intervals in air are, etc.[89]

By combining all formulations Y(1) through Y(6) we can establish the principal hypotheses of Young's theory of interference of light: 1) light of each color corresponds to a specific interval of periodicity in air ("hypothesis of periodicity;" Young's "interval" being equal to one half of the wavelength); 2) the velocity of light is inversely proportional to the index of refraction of the medum ("hypothesis of velocity"); 3) intersecting rays can reinforce or destroy one another ("hypothesis of interference"); 4) the maxima and minima occur where the path difference of intersecting rays contains, respectively, an even or an odd number of half-wavelengths ("the rule for locating maxima and minima," which is the result of applying the principle of superposition of waves to sinusoidal waves); 5) all phenomena of periodical colors can be described through interference of *two* rays, but the selection of proper pairs is specific for each phenomenon ("hypothesis of two-ray interference"); 6) interfering rays have the same frequency; 7) they originate from the same point; 8) they pass different routes and intersect again at a very small angle; 9) rays originating from natural light and polarised in perpendicular directions do not interfere; and 10) under some circumstances a reflected or diffracted ray may change its phase ("hypothesis of phase change").

Young certainly relied on the concept of waves to derive some of these hypotheses (the first two were, in fact, borrowed from the old wave theories). However, after obtaining them he reformulated them on the basis of rays of light so that he could apply his theory of interference without any regard to the nature of light. This does not mean that he abandoned the wave theory: he simply separated the defense of his principle of interference from that of the wave theory. To demonstrate the truth of the principle he had to show that the theory of interference could explain all phenomena of periodical colors. This, of course, also supported the wave theory, but in Young's eyes it was insufficient, for there remained other optical phenomena unexplained (dispersion, polarization, double refraction, etc.). At times, Young doubted that the wave theory will soon become powerful enough to account for all of them. However, despite this pessimism, he always argued in favor of the wave theory.[90]

89 Young, "Chromatics," *Encyclopaedia Britannica. Supplement to the 4th edition*, (Edinburgh, 1824), 3: 141–63. The half-volume with the article was published early in 1818. Here the references are given to *Misc. Works* 1:287, italics added.

90 See, for instance, Young, "Experiments and calculations," 188; *Lectures* 1:471, 482; "Chromatics," 326–40.

When searching for the *common* elements in formulations Y(1) through Y(6), one will find the hypothesis of periodicity, the hypothesis of two-ray interference, the rule for locating maxima and minima, and the conditions of coherence: this will constitute the content of Young's principle of interference of light. Thus, this principle is a core of Young's theory of interference. It will be shown further (see Ch. V) that the principle of interference in itself is sufficient for explaining quantitatively some phenomena of interference of light. For other phenomena, Young supplemented the principle of interference with other hypotheses from our list.

From this point of view, the discovery made in May 1801 probably consisted of a quantitative explanation of the colors of thin films, compatible with Newton's observations. By calling this theory a "law" Young probably wanted to emphasice that it was proven experimentally.

Chapter V

Young's theory of interference and its application

Most historians have believed that Young's case for the principle of interference of light was strong.[1] A few scholars, however, have concluded that his observations were too crude, some of his explanations were erroneous, and others overburdened with hypotheses.[2] Since neither of these opinions seems well documented, the question remains open. I intend to defend another view of Young's theory of periodical colors, which I will call the "theory of interference." It will be shown that, 1) Young constructed the theory of interference as a mathematical theory; 2) to support his theory, he presented observations of his own and of others, of such number and precision as he considered sufficient for periodical colors; 3) all his mathematical derivations and all his experiments (except for one) were accurate; 4) the presentation of his theory lacked mathematical details and a discussion of fundamental physical concepts; and 5) some of Young's alleged "mistakes" were caused by a miscomprehension of his works. My discussion will be based on a study of all of Young's works on interference of light written between 1801 and 1817.

A peculiar feature of Young's style was presenting only the results of his research with some hints for obtaining them, while omitting all physical reasonings and mathematical derivations. He set forth all of his principal optical hypotheses in his early papers of 1801 to 1803. When Young returned to a topic in his later works, he never expicitly stated whether he was introducing any theoretical innovations into it. This creates considerable difficulties in establishing and dating his ideas. When Young did not present any numerical data I was concerned with approximating his original formulae only insofar as to have them agreed

---

[1] See, for instance, Whewell, *History*, 3: 94–95; Frankel, "Jean-Baptist Biot," 300; and Silliman, "Augustin Fresnel," 110.

[2] Worrall, "Thomas Young," 144–55; and Latchford, "Thomas Young," 175–231. Some authors have attempted to distinguish the experimental foundation of the principle of interference from that of the wave theory of light. Morse, for instance, stated that the former "had been demonstrated and confirmed by unequivocal experiments" ("Natural philosophy," 134), while "Young did not found his theory on experiment or observations; he did found his hypotheses on other men's hypotheses" (ibid., 111). In Morse's view, the major flaw of Young's theory of light was the hypothesis of the luminiferous ether. Young has been criticized for applying interference to such phenomena of colors as the blackness of bodies ("Theory of light," 163–64), colors by refraction, and colors of flames ("Production of colours," 175–77). After 1802, however, Young employed this principle only to periodic colors. His theory of diffraction has also been criticized; see Whittaker, *History*, 103; Peacock, *Thomas Young*, 164; Morse, "Young, Thomas," 566; and Latchford, "Thomas Young," 229.

qualitatively with his explanations; but when numerical data were given, I attempted to derive exact equations which agree, as much as possible, with his data.

Since Young's explanations of different classes of phenomena are independent of each other, I have selected a topical arrangement of the discussion, which is supplemented by a chronological one within each class of phenomena.

Section I: Interference of reflected and refracted light

V.1 The colors of thin films

In May 1801 Young first applied the principle of interference to the colors of thin films.[3] The selection of interfering rays in this case was probably relatively simple. For the colors seen in reflected light, he chose those rays which were reflected once at each of the two parallel surfaces. He disregarded other rays which arrived in the same direction after being reflected several times at both surfaces, since they did not affect the path difference – Young's only concern. He calculated the path difference P of interfering rays SCO and $SAEFO$ (see Young's diagram in Fig. 19) as

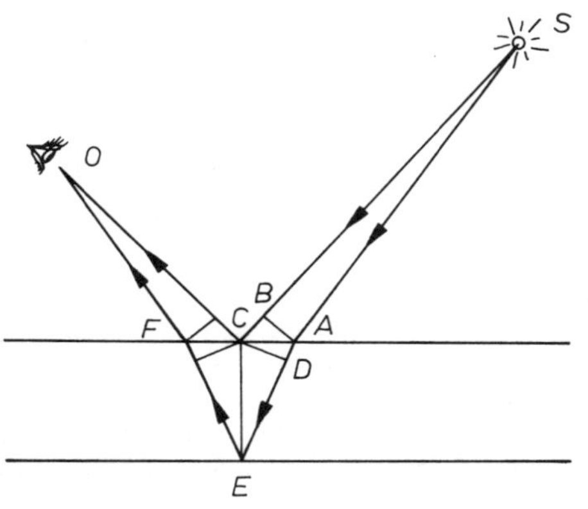

Fig. 19
Young's explanation of the colors of thin films (from *Misc. Works* 1: Fig. 130)

[3] Young, "Theory of light," 160–3.

$$P = 2DE = 2e \cos r, \tag{5.1}$$

where *AB* is perpendicular to *SC* and *CD* to *AE*, $e = CE$ is the thickness of the plate of air, and *r* is the angle of refraction. Young's derivation, with the aid of his diagram, is quite clear, though unusual for a modern reader.[4] To test this result, Young transformed it:

> Hence, that *DE* may be constant, or that the same color may be reflected, the thickness *CE* must vary as the secant of the angle of refraction *CED*; which agrees exactly with Newton's experiments; for the correction is perfectly inconsiderable.[5]

Thus, Young verified that for a constant path difference (or for the same color)

$$e \cos r = \text{const.} \tag{5.2}$$

This is a particular case of the more general modern equation

$$2ne \cos r = (2m - 1)\frac{\lambda}{2}. \tag{5.3}$$

where *m* is an integer, and λ is the wavelength. It will be shown later that Young knew this equation as well.

Some historians have considered Young's explanation of the colors of thin films to be merely an alternative interpretation of Newton's results.[6] In fact, there is a difference between Young's equation (5.2), which he derived theoretically, and Newton's empirical equation

$$e_r = \frac{e}{\cos U}. \tag{5.4}$$

where e_r and *e* are the thicknesses of the plate which reflects a given color when the angle of refraction is *r* or *0*, respectively, and *U* is an auxiliary angle defined as

$$\sin U = \frac{(105 + n)}{106} \sin r, \tag{5.5}$$

[4] Young later presented a derivation similar to the modern one; see his "Review of Malus," 270–71. He also discussed the fringes seen in transmitted light in "Chromatics," 290.

[5] Young, "Theory of light," 160.

[6] Latchford, "Thomas Young," 175; and Worrall, "Thomas Young," 149.

where n is the index of refraction.[7] When $r < 60°$, $\cos U \simeq \cos r$, and Newton's equation coincides with Young's. However, for larger angles of refraction the difference becomes significant (for instance, 10% for $r = 80°$).[8] Young chose to neglect this discrepancy, since he trusted both his principle and Newton's experimental skill. Fresnel suspected that Newton's observations at very large angles were erroneous, but in 1827 John Herschel still believed that Newton was right.[9] Only in 1849 was it found that Young's equation was correct and that Newton's results were inexact.[10]

According to Young's theory, the center of Newton's rings in reflected light should be white because the path difference is zero, but it is actually black. To explain this discrepancy he assumed that when light is propagated from a rarer to a denser medium and is reflected at the boundary, it changes to the opposite phase.[11] To verify his hypothesis of a phase change, in 1801 Young proposed an experiment:

> if a thin plate be interposed between a rarer and a denser medium, the colours by reflection and transmission may be expected to change places.[12]

Young omitted an important detail here, namely, the refraction of the thin film. In 1802, he observed that by interposing a drop of oil of sassafrass between a prism of flint glass and a lens of crown glass, he transformed the central black spot into a white one.[13] Young remarked that the index of refraction of this oil is intermediate to those of crown and flint glass. This interesting experiment is quite convincing; however, Young failed to mention that while interposing the drop of oil he kept the same pressure on the two glasses. By reducing the pressure he could have increased the distance between the glasses, and a white spot would have appeared instead of the black one. Brougham invoked this argument to demonstrate the innaccuracy of Young's observation.[14] In his reply to Brougham, Young remarked that he did not change the pressure.[15]

---

[7] Newton, *Opticks*, 204.
[8] Ibid.
[9] Fresnel, *Oeuvres complètes*, 3 vols. (Paris, 1866–70), 1: 57–58; hereafter cited as *Oeuvres de Fresnel* or *Oeuvres*. John Herschel, "Light [1827]," *Encyclopaedia Metropolitana* 4 (1848): 464–65.
[10] Frederic de La Provostaye, and Paul Desains, "Mémoire sur les anneaux colores de Newton," *Annales de Chimie* 27 (1849): 423–39.
[11] Young, "Theory of light," 154–55, 160.
[12] Ibid., 161.
[13] Young, "Production of colours," 175.
[14] [Brougham], "Young on colours not hitherto described," *The Edinburgh Review* 1 (1803): 458.
[15] Young, "Reply," 208.

It has not been previously noticed by historians that Young in fact used his hypothesis of phase change in the "Theory of light," *before* he confirmed it experimentally. I concluded this from a study of the table of the wavelengths, which Young claimed he derived from Newton's measurements of the thicknesses of a plate reflecting different colors (Table 2).[16] Using Newton's data I calculated the wavelengths, with the hypothesis of phase change and without it (see Table 1). In the former case we have (at normal incidence)

$$2\,e_m + \frac{\lambda}{2} = m\lambda, \tag{5.6}$$

where e_m is the thickness corresponding to the fringe of mth order. In the latter case we have

$$2\,e_m = m\lambda. \tag{5.7}$$

The first four columns in Table 1 present the wavelengths calculated by eq. (5.6) for the rings of the first four orders; the fifth column was calculated for the second order by eq. (5.7). It is clear (see Table 2) that Young used the hypothesis of phase change.

Two other points are worth noting here. First, the wavelengths derived from the fringes of the first order are too long. Young would not admit that Newton's measurements were mistaken, and he suggested that the discrepancy was due to ethereal atmosphere.[17] Second, all of Young's results are slightly higher than they should be, according to eq. (5.6). It is not clear why he modified these values. Perhaps he used some measurements of his own. It should be noted that Biot and Fresnel also calculated the wavelengths from Newton's data, and they also made some modifications (Table 2).[18]

V.2 The colors of "thick plates"

The first explanation of the colors of thick plates which was based on interference appeared in Young's "Theory of light."[19] Following Newton, Young attributes the major role in this phenomenon to the light scattered by tiny particles of dust (or of some other substance) at the first surface of a glass mirror. Neither this

[16] Young, "Theory of light," 161. Newton's results are taken from his *Opticks*, 233.
[17] Young, "Theory of light," 161.
[18] Biot, *Traité de physique*, 109; Fresnel, "De la lumière," *Oeuvres* 2: 24. Fresnel probably borrowed some of his wavelengths from Biot.
[19] Young, "Theory of light," 163.

account nor his later versions in his *Lectures* and "Chromatics" contained any diagrams or mathematical derivations.[20] Young may be interpreted in the following way. Let the ray *ABCM* (see Fig. 20), which is scattered by the particle *A* and

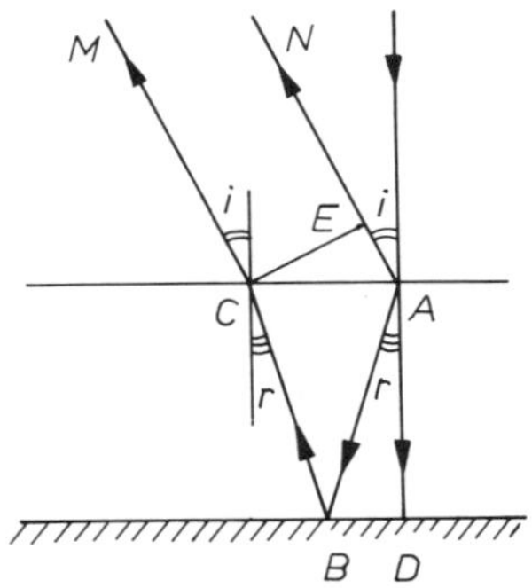

Fig. 20
An illustration to Young's explanation of the colors of "thick plates"

subsequently reflected at *B* and refracted at *C*, interfere with the ray *ADAN*, which first passes by the particle *A* and then returns after reflection at *D* to be scattered at *A*. Let the emerging rays be parallel. It follows from the diagram that path difference $P=(AB+BC)n-(2ADn+AE)$. Since $AB=BC=e/\cos r$, and $AE=2e(\tan r)(\sin i)$ (where *e* is the mirror's thickness, and *i* and *r* are the angles of incidence and refraction, respectively) we have

$$P = 2ne\,(1-\cos r) = 4ne\sin^2\frac{r}{2} = \frac{e\tan^2 i}{n}. \tag{5.8}$$

Since the colored rings can be seen only at very small angles of refraction, when $\sin i \simeq \tan i \simeq i$, the equation for bright fringes becomes

$$\frac{e\varrho^2}{nR^2} = m\lambda, \tag{5.9}$$

or

$$\varrho = \sqrt{\frac{mn\lambda}{e}} \times R, \tag{5.10}$$

[20] Young, *Lectures* 1:471, and "Chromatics," 300–301.

where ρ is the ring's radius, R is the distance from the screen to the plate, λ is the wavelength, and m is an integer.

Young's derivation might not have been exactly like this one, but his final result had to be similar, for he claimed that it agreed with Newton's observations, and, according to Biot, Newton had utilized equation (5.10).[21] In 1827, John Herschel, probably influenced by Young, arrived at the same equation in a way slightly different from the one presented above.[22] George Stokes (1819–1903), the author of the best mathematical account of the colors of thick plates, gave due credit to his predecessor:

> Dr. Young's explanation is however excessively brief; and he has rather pointed out the application of the grand and newly-discovered principle of interference to the explanation of the phenomenon, than followed the subject into any of its details. At the same time, it appears evident, from an attentive perusal of what he has written, that at least the broad outlines of the complete explanation were clearly presented to his mind.[23]

In 1807, William Herschel (1738–1822) observed colored rings by scattering a fine powder into a beam of light falling on a concave metallic mirror.[24] He believed that this phenomenon refuted Newton's theory of fits, which required two surfaces to form the colors, while there was only one in Herschel's experiment. In 1809, in his review of Herschel's paper, Young hinted that Herschel's colors had the same origin as those observed by Newton and Michel d'Ailly, Duc de Chaulnes (1714–1769).[25] In 1817 he presented a more detailed explanation of all these phenomena, based on the interference of light scattered by a *single* particle placed before a reflecting mirror.[26] However, even this "detailed" account is hardly comprehensible, and there is no evidence that before 1827 anyone understood Young's explanation of the colors of thick plates and related phenomena.[27]

---

21 Biot, *Traité de physique* 4: 166.

22 Herschel, "Light," 473–74.

23 Stokes, "On the colours of thick plates," in *Mathematical and Physical Papers*, 5 vols, 2nd ed. (New York, 1966), 3: 156.

24 Herschel, "Experiments for investigating the cause of the coloured concentric rings, discovered by Sir Isaac Newton, between two object glasses laid upon one another," *Phil. Tans.*, 1807: 231–2.

25 [Young], "Herschel on concentric rings between two object-glasses," *Retrospect of Philosphical . . . Discoveries* 4 (1809): 24. The identification of the author is mine. It is based on implicit references to Young's works on the production of colors of thin films with the use of a solar microscope, on halos, on the difference between the colors of thin films, thick plates, and Herschel's colors.

26 Young, "Chromatics," 300–301.

27 Besides John Herschel, in 1827 Jacques Babinet presented a theory of thick plates.

V.3 The colors of the "mixed plates"

In the spring of 1802, Young discovered a new variety of colors of thin films which he called the "colors of mixed plates".[28] This phenomenon is observed when the space between two plates of glass is filled with small portions of two different substances, for instance, air and water. In this case Young assumed that each of the two interfering rays travels the same distance but through only one of the substances, so that their path difference is due to a difference in their velocites (or in their indexes of refraction). From this hypothesis he calculated that for a mixture of air and water (with a ratio of velocities of 4 to 3) the fringes in mixed plates must appear at a thickness six times greater than that of a plate of air producing fringes of the same order.[29] His experiment agreed with this prediction, thus confirming that light travels faster in a rarer medium. Young never published his derivation of the path difference in mixed plates, but Emile Verdet reconstructed it and found it to be in a complete agreement with Young's results.[30] Young probably soon realized that this explanation was too concise, for when reprinting his paper in 1807 in his *Lectures*, he added some details and a diagram.[31] Unfortunately, this diagram (see Fig. 21) was too complicated to clar-

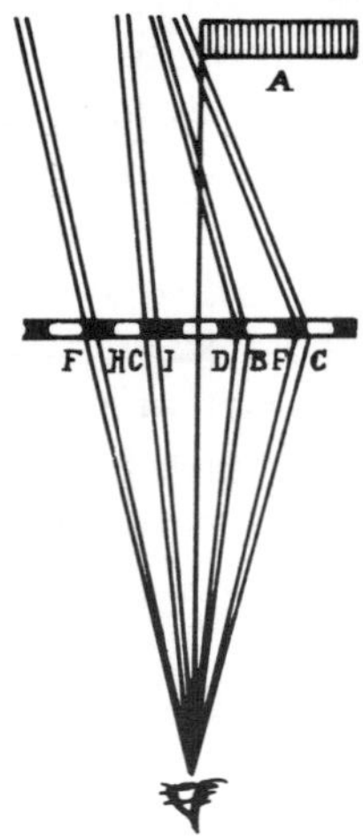

Fig. 21
Young's illustration of the colors of "mixed plates" (from his *Lectures* 2: Fig. 112). Black and white spaces represent air and water.

[28] T. Young's lecture notebooks, vol. 16, fol. 16L (when a microfilm frame contains two pages L and R mean "left" and "right," respectively): University College, London, Ms. Add. 13.

[29] Young, "Production of colours," 175.

[30] Verdet, "Leçons d'optique physique," 2 vols. in *Oeuvres de E. Verdet*, 8 vols. (Paris, 1869–72), vols. 5 & 6, 1: 155–59.

[31] Young, *Lectures* 2: 636, Plate 15, Fig. 112.

ify the calculation of the path diference. Young's account of the "mixed plates" in the "Chromatics" was even less clear than his earlier ones.[32]

In 1814, Young applied the principal idea of his explanation of "mixed plates" to the phenomenon of "chromatic polarisation." In 1811, Arago discovered that polarized light exhibited colors when it passed through thin plates of some doubly refracting substances.[33] In 1812–13, Biot thoroughly investigated these colors and formulated several phenomenological laws on how they depend on the crystal's thickness and its orientation to the incident light.[34] He found, for instance, that the colors were the same as the colors of thin films whose thickness was 260 times less than that of the crystal. Young discovered in 1814 that all of Biot's laws can be obtained from his principle of interference by assuming that the colors result from the interference of the ordinary and extraordinary rays propagating in the same plate.[35] He said that these colors

> are, in fact, merely varieties of the colours of 'mixed plates,' in which the appearances are found to resemble the colours of simple thin plates, when the thickness is increased in the same proportion, as the difference of the refractive densities is less than twice the whole density.[36]

He gave the numerical example: $n_e/n_o = 160/159$ then $e_d/e_s = 318$, where n_e and n_o are the indexes of refraction of the extraordinary and ordinary rays, respectively, and e_d and e_s ar the thicknesses of the doubly- and singly-refracting substances, respectively. The equation for the bright fringes in the crystal is:

$$(n_e - n_o)\, e_d = m\lambda, \tag{5.11}$$

Where λ is the wavelength, and m is an integer. The analogous equation for the singly-refracting plate (in transmitted light) is

$$2n_o e_s = m\lambda. \tag{5.12}$$

From these two equations:

$$\frac{e_d}{e_s} = \frac{2\,n_o}{n_e - n_o} \tag{5.13}$$

---

[32] Young, "Chromatics," 307–308.

[33] Arago, "Mémoire sur une modification remarquable qu'éprovent les rayons lumineux dans leur passage a travers certain corps diaphanes d'optique," *Mem. Acad.*, 1811: 93–134.

[34] Biot, "Mémoire sur un nouveaux genre d'oscillation que les molécules de la lumière éprouvent en traversant certains cristaux," *Mém. Acad.*, 1812: 1–371.

[35] Young, "Review of Malus," 266–73.

[36] Ibid., 269.

It is easy to see that this equation agrees with Young's statement and his numerical example. Thus, it seems that Young applied eq. (5.11).

Young believed that the difference between his theoretical prediction (318) and the experimental result (360) was sufficiently small, because

> The measures which Mr. Biot has obtained differ much less from the calculation derived from these principles only, than they differ among themselves; and we cannot help thinking such a coincidence sufficient to remove all doubts (if any existed) of the universality of the law on which that calculation is founded; . . . an error of a single unit in the third place of decimals of the index of refractive density, as determined by Mr. Malus, would be sufficient to make the coincidence perfect: and a greater degree of accuracy can scarcely be expected in experiments of this kind.[37]

Fresnel afterwards constructed a more exact theory of chromatic polarization, but its core – the interference of two polarized rays propagated in a crystal with different velocities – remained the same as in Young's theory.[38]

V.4 The colors of supernumerary rainbows

Supernumerary rainbows are the colored fringes seen sometimes inside the primary and outside the secondary bow. Young's interest in this phenomenon was probably stimulated by Henry Pemberton (1694–1771), Professor of Medicine at Gresham College in London, who in 1722 gave the first explanation of it by the theory of fits.[39] In 1803, Young probably already possessed an explanation of this phenomenon, though he presented only some conclusions about the conditions for observing it. He found theoretically that distinct supernumary bows could be observed only if a large number of rain drops were of the same size; for instance, a distinct red fringe would be seen at the distance of 2° from the red of the primary rainbow when the drops' diameters were between 1/70 and 1/80 in.[40] There

[37] Ibid., 269–70.

[38] Fresnel, "Note sur le calcul de teintes que la polarisation développe dans les lames cristallisées," *Oeuvres* 1:609–53.

[39] Pemberton, "A letter . . . concerning the abovementioned Appearance in the Rainbow," *Phil. Trans.*, 32 (1722): 245–48.

[40] Young, "Experiments and calculations relative to physical optics [1803]," *Misc. Works* 1: 185–86.

were no futher details in the *Lectures*.[41] In 1817, Young gave a fuller account of the phenomenon.[42] He said that the bow of the mth order is produced by rays reflected within the drop m times. He also presented (without a derivation) the path difference for interfering rays. It is possible to reconstruct Young's derivation.

Let the incident rays RA and QB be parallel to the diameter OC of a drop (see my Fig. 22). The points of incidence A and B are chosen so that the refracted rays

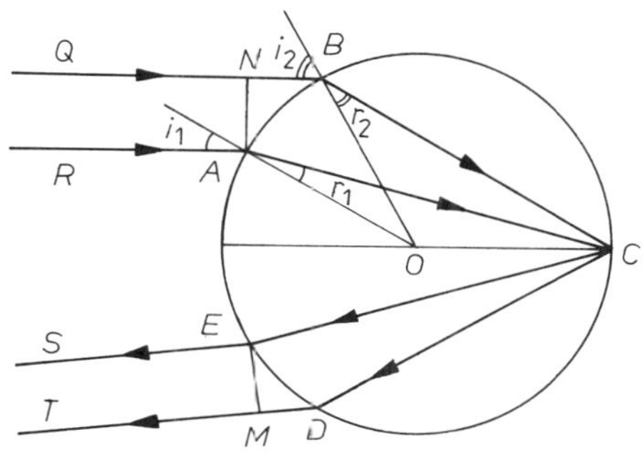

Fig. 22
An illustration to Young's explanation of the supernumerary rainbows

meet at the point C. After being reflected there and then being refracted at D and E, these two rays again become parallel. The path difference of the interfering rays is

$$P = n\,(AC + CE - BC - CD) - DM - BN, \tag{5.14}$$

where AN is perpendicular to QB and EM to DT, and n is the index of refraction of the drop. Since $AC = CE$, $BC = CD$, and $BN = DM$, we have

$$P = 2\,(AC - BC)\,n - 2BN$$

or

$$P = 2R\,[n\,(\cos r_1 - \cos r_2) - (\cos i_1 - \cos i_2)], \tag{5.15}$$

where R is the drop's radius, i_1 and i_2 are the angles of incidence of the two rays, and r_1 and r_2 are their angles of refraction. This equation agrees with Young's result expressed in non-mathematical form.[43] From this formula Young derived the angular distances of the supernumerary bows from the primary one. Since he did not know the size of the drops, he could not verify his

[41] Young, *Lectures* 1:470–71.
[42] Young, "Chromatics," 293–97.
[43] Ibid., 295.

equation directly. Instead, he calculated from it the diameter of drops which would produce a particular kind of the rainbow. It was later shown that Young's explanation of supernumerary bows was a first approximation to a more general theory of the rainbow.[44] There is no evidence that anyone had grasped Young's terse explanation of this phenomenon during his lifetime.

Section II: Interference of diffracted light

V.5 The colors of striated surfaces

In the "Theory of light" Young's account of the colors produced by scratches on polished surfaces appears as his first application of the principle of interference, although it was actually worked out later than those of the colors of thin films, thick plates, and diffraction.[45] He probably considered this to be the simplest illustration of his principle. To study this phenomenon Young utilizied the scale of a micrometer, since the regularly arranged lines provided a much better opportunity for a quantitative study of the phenomenon than an isolated scratch. He used a Coventry micrometer, a glass plate with a series of parallel lines 1/500 in apart. Some of these lines turnes out to be doublets with a separation about 1/10,000 in.

Using this grating, Young made the following observation:

> I placed one of these so as to reflect the sun's light at an angle of 45°, and fixed it in such a manner, that while it revolved round one of the lines as an axis, I could measure its angular motion; and I found that the brightest red colour occured at the inclinations 10.5°, 20.75°, 32°, and 45°; of which the sines are as the numbers 1, 2, 3, and 4.[46]

---

[44] In 1835 Richard Potter independently developed a theory similar to Young's (see his "Mathematical considerations on the problem of the rainbow, shewing it to belong to physical optics," *Trans. Cambr. Phil. Soc.* 6 (1838): 141–52), and arrived at the same results. Airy then developed a more general theory on the basis of the Huygens-Fresnel principle and found the fringes shifted relative to Young's particularly in the first order. See George Airy, "On the intensity of light in the neighbourhood of a caustic," *Trans. Cambr. Phil. Soc.* 6 (1838): 391–92. See also Carl Boyer, *The Rainbow from Myth to Mathematics* (New York, 1959), 294–304.

[45] Young, "Theory of light," 158–60. This observation is the only one of those presented in this paper which was not mentioned in Young's letter to Nicholson (see "Letter . . . respecting sound and light," 132) dated July 13, 1801. Young did not consider this to be a diffraction phenomenon.

[46] Ibid., 159. The precision here is about 10% $(45° - 4 \times 10.25°)/45°$.

He assumes that each point of a scratch reflects light in all directions, and those rays interfere which are reflected by different scratches in the same direction. Young's derivation of the path difference differs only slightly from the modern one. He considers the path difference of rays reflected from two points of the adjacent scratches A and B (see my Fig. 23) when both points lie

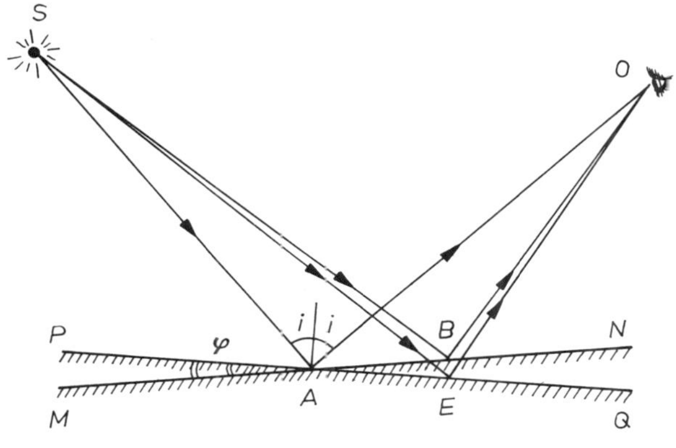

Fig. 23
Young's illustration of his explanation of the colors of striated surfaces (a slightly modified diagram from *Misc. Works* 1: Fig. 129)

in a given plane MN. The initial position of this plane is taken so that the sun's image appears to coincide with points A und B. In this case the routes of the two reflected rays are nearly equal, or $SA+AO=SB+BO$. When MN is rotated into the position PQ, the path difference becomes $P=(SE+EO)-(SA+AO)=(SE+EO)-(SB+BO)$, where E is the new position of B. Since BE is approximately perpendicular to MN, then, according to Young, $P=BE\cos i$, where i is the angle of incidence, and BE, the depression of the point B, which is a constant in this experiment. Young says that the depression is proportional to $\sin\varphi$ (more exactly, $BE=d\sin\varphi$, where φ is the angle of rotation of the plane, and d is the distance between the two ajacent scratches). Therefore, the equation for the bright fringes is

$$2d\sin\varphi\cos i=m\lambda, \tag{5.16}$$

where m is an integer, and λ the wavelength. Since i, d, and λ are constant, then

$$\sin\varphi_m/m=\text{const}, \tag{5.17}$$

which is precisely what Young derived.

Although Young knew the distance *d* and the wavelength λ, he never discussed equation (5.16). It seems he had difficulties in identifying the constant *d*. As mentioned above, Young's grating had two constants of periodicity, and intially he was not sure whether the interfering rays came from two components of the same scratch or from different scratches. In 1813, he realized that the colors observed were due to the smaller distance.[47] Indeed, taking $\lambda = 2.32 \times 10^{-5}$ in for red light, as Young did, I found by eq. (5.16) $d = 0{,}9 \times 10^{-4}$ in, which is only 10% less than Young's $d = 1.0 \times 10^{-4}$ in, a precision, which he considered to be satisfactory in investigations of periodic colors. Young also says that he acquired another Coventry micrometer where all the scratches were singlets, and that observations with this micrometer fully confirmed the theory.[48] Unfortunately, he gives no information about the angular positions of the fringes, and it is impossible to check his conclusion.

Young utilized only one pair of adjacent scratches to calculate the path difference, since he believed that all other scratches simply added to the intensity of the fringes. In 1814, David Brewster (1781–1868) discovered that mother-of-pearl and other naturally striated substances produce unusually pure colors separated by dark intervals. To explain these colors Young had to revise his model.[49] In 1817, he presented an explanation of Brewster's colors, which contained the first sketch of the theory of the diffraction grating:

> This circumstance may be satisfactorily deduced from the general law, if we consider that each interference depends not only on two portions separated by a simple interval, but also a number of other neighbouring portions, separated by other intervals which are its multiples; so that unless the difference of two paths agrees very exactly with the interval appropriate to each ray, the excess or defect being multiplied in the repetitions, the color will disappear; consequently, each of the stripes which, in other cases, divide the space in which they appear almost equally between light and darkness, when homogenous light is employed, becomes here a narrow line, and their succession affords a spectrum exhibiting very little mixture of the neighbouring colours with each other, and nearly resembling that which is afforded by the simple dispersion of the prism.[50]

[47] Young, "Remarks on the measurement of minute particles, especially those of the blood and of pus [1813]," *Misc. Works* 1:356.
[48] Ibid.
[49] Brewster, "On the new properties of light exhibited in the optical phenomena of mother-of-pearl, and other bodies to which the superficial structure of that substance can be communicated," *Phil. Trans.*, 1814:397–418.
[50] Young, "Chromatics," 299.

There is some possibility of confusion here because Young uses the same term "interval" for the spacing between two adjacent lines and also for the wavelength. I believe that his expression "the interval appropriate to each ray" refers to the latter, as we can see from the following illustration of Young's idea. Let us assume that the green region in the solar spectrum extrends from 500 nm to 580 nm. Then, let us imagine a grating illuminated by monochromatic light of 540 nm and forming the mth maximum at the angle Θ. Now let us replace the monochromatic light with sunlight and examine what happens to light of say about 520 or 560 nm diffracted at the same angle Θ. If the number of lines in the grating is large enough, those rays which are reflected by every thirteenth line will acquire a path difference of about one-half of the wavelength of 540 nm. Thus, the regions of the spectrum adjacent to the wavelength of 520 nm and 560 nm will be extiguished, and the green line in the diffraction spectrum will be much narrower than in the solar spectrum. Moreover, there will be dark spaces between the green and yellow fringes, or between the green and blue ones, with practically no mixture of colors.

V.6 Diffraction of light by a narrow body: internal fringes

In 1801, in his "Theory of light" Young presented the first explanation of diffraction of light by his principle of interference.[51] As in his earlier "Outlines of experiments," he continues to treat diffraction as a refraction in an ethereal atmosphere surrounding bodies, but this atmosphere is now unlimited and has a density gradient. This change was probably affected by his study of refraction in a medium with a variable optical density, which he applied in 1800 to atmospheric refraction and to the human eye.[52] In his new model rays passing at different distances from the body (see Fig. 24) are unequally deflected from their initial direction. Some of them interfere and produce the internal fringes. Young does not explain how he derived the equation for the path difference; he only remarks that his theory agrees with the observations of Newton and later authors.[53]

By 1802, Young became aware of several physical and mathematical arguments opposed to this ethereal model. First, the density gradient could be determined only empirically, by taking the principle of interference for granted. Secondly, observations of the independence of diffraction of the substance of

[51] Young, "Theory of light," 164–66.
[52] Young, "Mechanism of the eye," 20–21; and "Lettre ... sur une formule pour les réfractions," 360–63.
[53] Young, "Theory of light," 165.

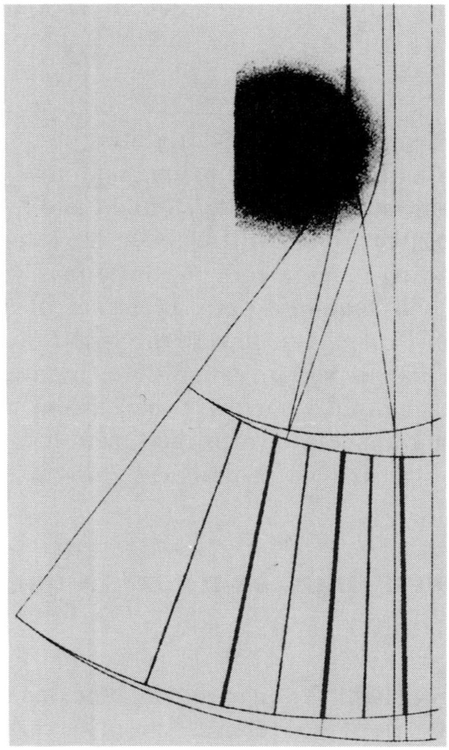

Fig. 24
Young's illustration of his first theory of diffraction (from *Misc. Works* 1: Fig. 131)

a diffracting body turned Young's attention to the hypothesis of Grimalid and Hooke that diffraction is a natural tendency of light to diverge near a body.[54]

[54] Young said in 1799 that no *"comparative* experiments have been made on the inflection of light by substances possessed of *different refractive powers*" ("Outlines of experiments," 83, italics added). Grimaldi and Newton observed qualitatively that diffractors made of various substances produced fringes; in particular, Newton experimented with glass. Since Young obviously knew about Newton's observations, he possibly meant *quantitative* experiments. In 1801 Young noted that he would not consider his hypothesis of the ethereal atmosphere certain, "until *further* experiments have been made on the inflecting *power* of different substances" ("Theory of light," 165, italics added). Finally, in 1803, while describing his studies of diffraction, Young concluded: "I have not, *in the course of these investigations* found any reason to suppose the presence of such an inflecting medium in the neighbourhood of dense substances as I formerly inclined to attribute to them" ("Experiments and calculations," 188, italics added). I am not aware of any of Young's contemporaries who made public how fringes' positions depend on the substance of a diffractor, and because of that I interpret the cited passages as evidence that between 1801 and 1803 Young made such observations himself.

Thirdly, in the spring of 1802 he found it difficult to reconcile his ethereal model with the accepted explanation of stellar aberration.[55] By that time he conceded that his hypothesis of the ethereal atmosphere "is not absolutely necessary."[56] On July 1, 1802, he presented a new theory of diffraction in his paper "Production of colours."

His new explanation assumed that light has a natural tendency to bend into a shadow. When viewing a distant source of light with a narrow body (a wire or a hair) held near his eye, Young noticed colored fringes surrounding the luminous body. He supposed that,

> their cause must be sought in the interference of two portions of light, one reflected from the fibre, the other bending round its opposite side, and at last coinciding nearly in direction with the former portion . . . It was easy to calculate, that for the light least inflected, the difference of the paths would be to the diameter of the fibre, very nearly as the deviation of the ray, at any point, from the rectilinear direction, to its distance from the fibre.[57]

To measure the angular distance of the first bright fringe from the body, Young took a card with a rectangular hole in it and attached a hair in the middle of it. Looking through this hole at the flame of a distant candle, he counted the number of fringes inside the hole. Young's equation for the first maximum according to the passage just cited is

$$P/d = x/b, \tag{5.18}$$

where $P = AC$ is the path difference (see my Fig. 25), $d = AB$ is the diameter of the fiber, $x = MN$ ist the "deviation" of the ray, and $b = BN$ ist the distance from the fiber to the eye. Let us now reconstruct the path difference as the difference of the routes AM and BM. Since

$$BM = \sqrt{b^2 + x^2} \simeq b + \frac{x^2}{2b},$$

and

$$AM = \sqrt{b^2(x+d)^2} \simeq b + \frac{x^2}{2b} + \frac{xd}{b},$$

we have

$$P = \frac{xd}{b}, \tag{5.19}$$

[55] Young's notebooks, vol. 16, fol. 20L.
[56] Ibid., fol. 20R.
[57] Young, "Production of colours," 171.

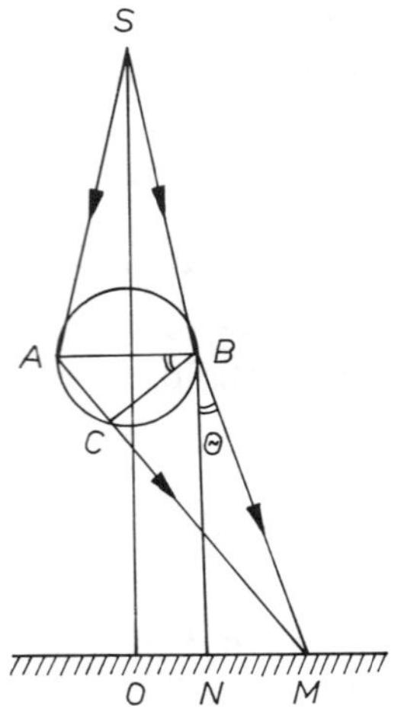

Fig. 25
An illustration to Young's theory of diffraction (the internal fringes)

which is the same as Young's result. Although Young described his formula only for the first maximum ($p=\lambda$), it can be used for a fringe of an arbitrary order (Young discussed an increase of the path difference with the increase of the angle of diffraction.) If the angle of deviation of the fringe is Θ, which usually is small, we have $\sin\theta \simeq \tan\theta = x/b$, and the angular size of the bright fringe of mth order is

$$\tan\theta_m = m\frac{\lambda}{d}, \tag{5.20}$$

Thus, Young discovered that the angular dimension of a fringe of a given color and order depends only on the size of the body.[58] He generalized this result for a number of diffracting bodies:

> When a number of fibres of the same kind, for instance, a uniform lock of wool, are held near to the eye, we see an appearance of halos surrounding a distant candle; but their brilliancy, and even their existence, depends on the uniformity of the dimensions of the fibres; and they are

[58] Young's hair experiment shows the limitation of the common definition of the internal and external fringes. It is more reasonable to distinguish them by the way of their production, that is whether they are formed by two rays inflected by the edges of a body or by a direct and an inflected rays.

> larger as the fibres are smaller. It is obvious that they are the immediate consequences of the coicidence of a number of fringes of the same size, which, as the fibres are arranged in all imaginable directions, must necessarily surround the luminous object at equal distances on all sides, and constitute circular fringes.
>
> There can be little doubt that the colored atmospherical halos are of the same kind: their appearance must depend on the existence of a number of particles of water, of equal dimensions, and in a proper position, with respect to the luminary and to the eye.[59]

Young possibly borrowed the idea that the size of a halo depends on the diameter of the water drops in a cloud from Newton.[60] Newton claimed that his explanation of halos, which he based on the theory of fits, agreed with his observations. There is, however, a considerable difference in the diameter of the water drops given by Newton himself and those derived from his observations by Biot, who reconstructed Newton's formula.[61] Biot considers the halos to be produced by rays scattered inside the drop. In Young's model, on the contrary, the rays do not enter the drops but are diffracted or reflected at their surfaces. Brougham believed this to be a flaw in Young's theory, since in his view, the colors were due to refraction.[62] Young certainly could not present any physical reasons for neglecting the refracted rays. His only argument could have been mathematical: the rays must originate at the edges to obtain a quantitative explanation of the phenomenon.

Assuming that Young applied eq.(5.20) to halos, let us see how well his theory agrees with measurments. In his *Lectures* Young claimed that to produce a series of halos, that are distant 2°–3° from one another, the diameter of a water drop must be greater than 1/2000 in.[63] This agrees with the size of drops calculated according eq. (5.20) from observations of Newton and Jordan (see Table 3).[64] Additional evidence of the use of eq.(5.20) can be found in Young's work with his *eriometer*. In 1813 he published in his *Introduction to Medical Literature* a description of a device called an "eriometer," to measure the size of particles of blood, threads of wool, and other small objects. The eriometer had a plate with a hole in it to view a distant source of light. The object was placed on a support between the hole and the source of light. To observe the colored rings around the

---

[59] See n. 57, 172.
[60] Newton, *Opticks*, 313.
[61] Biot, *Traité de physique* 4:242.
[62] Brougham, "Young on colours not hitherto described," *Edinburgh Review* 1 (1803), 457.
[63] Young, *Lectures* 1:466.
[64] Jordan, *An Account of the Irides or Coronae* . . . (London, 1799).

object there were many small holes along the circumference of the plate, with the principal one at the center. The object was moved about the plate until the colored rings produced by it became visible through the small holes. The eriometer turned out to be a simple and convenient instrument, and until recently it was used in medicine to measure the size of blood particles.[65]

It is quite clear from Young's description that the eriometer was based on his 1802 theory of internal fringes, and that he made quantitative observations of halos produced by small particles before 1813. One of his improvements in 1813 was a higher precision of experimental data. In 1813, Wollaston invented a micrometer for exact measurements of very thin wires.[66] He measured for Young several very small objects, and with these data Young found experimentally that

$$\tan \theta = \frac{K}{d}, \tag{5.21}$$

where Θ is the angular radius of the first ring, d is the diameter of the body, and K is a constant (1/30 000 in). Young remarks that,

> Some former investigations had led me to attribute to this unit a value somewhat smaller.[67]

In another section of the same book Young mentions K = 1/44 000 in, which is equal to the wavelength calculated by him in his 1802 hair experiment.[68] Thus, the earlier constant agrees with eq. (5.20). The new cosntant (1:30 000) may be introduced into the same equation by replacing $m = 1$ by $m = 3/2$ for the maximum of the first order. To justify this, Young assumed that inflected light has its phase inverted. With this hypothesis, we obtain instead of eq. (5.20)

$$\tan \theta_m = \left(m + \frac{1}{2}\right)\frac{\lambda}{d}. \tag{5.22}$$

[65] Young, "On blood and pus," 346–51; and Wood, "Thomas Young," 85. Privately, Young first described his "Agricultural Eriometer" to Joseph Banks, President of the Royal Society, as a simple and cheap device for farmers to measure the thickness of wool. He also presented Banks with a sample of his apparatus and a table of data for different brands of wool and a few other substances. See Young to J. Banks, September 10, 1810: British Library, Banks Papers, Add Ms 33, 982, ff.9–10.

[66] Wollaston, "Description of a single-lens micrometer," *Phil. Trans.*, 1813: 119–22.

[67] Young, "On blood and pus," 349.

[68] Ibid., 356. In Young's letter to Banks (see n.65), $K = 1/45\,000$ in.

In 1817, in his "Chromatics", Young gave the following numerical example. He said that particles of water 1/2185 in. diameter produce a ring of the angular radius $\theta = 4°$, and that $\tan\theta = 1/14$.[69] Since Young made his observations at the border of the green and red of the first order ($m = 1$, and $\lambda = 2.3 \times 10^{-5}$ in), we see that he applied eq. (5.22).

Thus, Young employed the same theory for colors produced by both small opaque bodies and water drops. In 1802, he used eq. (5.20), but in 1813 he introduced the hypothesis of phase change and arrived at eq. (5.22). The revision of the theory was due to new more precise observations. A later research had shown that eq. (5.22), which is probably the same as Young's, is a good approximation to the diffraction theory of halos.[70] During his lifetime, Young's thorough work on the quantitative explanation of internal fringes was ignored. The only observation of the internal fringes that attracted his contemporaries' attention was a qualitative experiment which I will call his "screening" experiment.

Some time in 1803, while observing the internal fringes produced by a narrow body *AB* (see my Fig. 26), Young took a little obstacle *PQ* and put it behind the

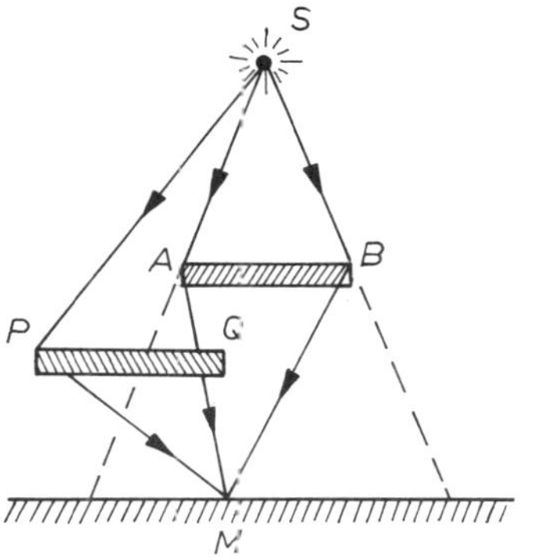

Fig. 26
An illustration of Young's "screening" experiment: AB is a narrow body (a wire), PQ is a screen to intercept light

[69] Young, "Chromatics," 305–306.

[70] Verdet found theoretically and confirmed experimentally that a number of irregularly distributed equal globules produce colored rings which correspond to m = 1.48 ; 2.4; 3.3; . . ., while Young's eq. (4.22) assigns to these rings m = 1.5 ; 2.5; 3.5; . . . Thus Young's explanation is valid, at least for fringes of low orders. See Verdet, "Sur l'explication du phénomène des couronnes," *Annales de Chimie* 34 (1852): 129–40.
High school teachers repeated at the Bakken Museum in Minneapolis Young's experiments with a hair and eriometer and obtained the same precision as Young.

body to intercept the light passing by one side of it. He discovered that all the internal fringes disappeared and concluded that

> these fringes were the joint effect of light passing on each side of the slip of card, and inflected, or rather diffracted, into the shadow.[71]

He proved that a decrease of the intensity of light due to the interception of one of the two beams of light was not responsible for the disappearance of the fringes, since even after reducing the intensity of both portions of light ten or twenty times he could see the fringes.

A closer examination shows, however, that Young's explanation of this experiment was not satisfactory. First, it is not clear whether the screen *PQ* really "intercepted" the ray *AM*. While passing very closely to the edge *Q* of the screen, this ray might have experienced another inflection at *Q* and come into the shadow again. To intercept the beam *AM* with certainty, Young could attach the screen *PQ* to *AB*, forming a large diffracting body (just as Fresnel did later; see Ch. VII). Secondly, even admitting that the ray *AM* was intercepted, Young had to explain why he did not take into consideration the ray *PM*, that originated from the ray *SP* inflected at the new "edge" *P* (since *PQ* is a "little screen," it must inflect light at both edges). Young says nothing about why this ray *PM* cannot interfere with *BM*. Thirdly, Young knew from his 1802 hair experiment that sometimes even a very narrow body does not produce internal fringes. In the experiment with a hair described above he found that when he replaced a human hair with a thicker horse hair, the fringes "were no longer visible."[72] Thus, although neither of the two portions of light was intercepted, the internal fringes disappeared anyhow. We will see later that in 1803 Young already knew that in some circumstances two portions of light cannot interfere. However, he carefully avoided mentioning such a possibility. Obviously, Young learned nothing from his "screening" experiment that he had not known before. It seems that he offered it as the simplest means to demonstrate the existence of interference of light. However dubious his reasoning, no one objected to it, and eventually this experiment was acknowledged to be a crucial argument in favor of the principle of interference (see Ch. VIII). It seems, however, that Young himself lost confidence in the persuasive power of the "screening" experiment, since he never mentioned it after 1803.

[71] Young, "Experiments and calculations," 179–80.
[72] Young, "Production of colours," 171.

V.7 Diffraction of light by a narrow body: external fringes

Young's first account of the external fringes appeared in 1803 in the "Experiments and calculations."[73] Here Young discusses diffraction by a slit formed by the edges of two knives (Newton's observation 9) and by a narrow body (Newton's observation 3 and two of Young's own experiments). In the former case, Young assumes that the dark fringe opposite the center of the slit is formed

> by the first interference of the light reflected from the edges of the knives, with the light passing in a straight line between them.[74]

According to this hypothesis, the path difference $P = SA + AO - SO$ (see my Fig. 27). Let a and b be the distances from the slit to the source of light and to the observation screen, respectively, and d the slit's width $(d \ll a,\ d \ll b)$. Then

$$SO = a + b, \quad SA = \sqrt{a^2 + \frac{d^2}{4}} \simeq a + \frac{d^2}{8a},$$

$$OA = \sqrt{b^2 + \frac{d^2}{4}} \simeq b + \frac{d^2}{8b},$$

and the path difference is

$$P = \frac{d^2 (a + b)}{8ab}. \qquad (5.23)$$

Since the "first interference" means $P = \lambda/2$, we obtain

$$\lambda = \frac{d^2 (a + b)}{4ab}. \qquad (5.24)$$

Using Newton's measurements in his Observation 9, I calculated the wavelength according to eq. (5.24) (see Table 4, column 4) and found them virtually identical with Young's result calculated with the same data (column 3).

---

[73] Young, "Experiments and calculations," 181–84.
[74] Ibid., 181.

For diffraction by a narrow body Young gives the wavelengths without any explanation, stating that they are the "results of a similar calculation."[75] I assumed that by "similarity" he means that in this case, as well as for a slit, the direct ray interferes with the reflected one, which would agree with his later statement in his *Lectures*.[76] To verify this interpretation I calculated the wavelengths for Young's measurements in his "Experiments and calculations" and compared them with his results. For an external fringe the path difference is $P = SA + AM - SM$ (see my Fig. 28). Let $MN = x$ be the distance of the fringe from the border of the geometric shadow of a body, and $AB = d$, the body's width.

Then

$$SA = \sqrt{a^2 + d^2 4} \simeq a + \frac{d^2}{8a},$$

$$AM = \sqrt{b^2 + \left(x + \frac{bd}{2a}\right)^2} \simeq b + \frac{x^2}{2b} + \frac{xd}{2a} + \frac{bd^2}{8a^2};$$

and

$$SM = \sqrt{(a+b)^2 + \left[x + \frac{(a+b)\,d}{2a}\right]^2} \simeq a + b + \frac{x^2}{2\,(a+b)} + \frac{xd}{2a} + \frac{d^2\,(a+b)}{8a^2}.$$

Finally

$$P = \frac{x^2 a}{2b\,(a+b)}. \tag{5.25}$$

According to Young, the second bright line corresponds to the path difference $P = \lambda$, and the second dark line to $P = 1.5\lambda$.[77] This implies that bright fringes are seen where $P = (m-1)\,\lambda$; while dark fringes correspond to $P = (2m-1)\,\lambda/2$ (in both cases $m = 1, 2, 3 \ldots$). The first bright fringe is located at the border of the

[75] Ibid.
[76] Young, *Lectures* 1:467.
[77] Young, "Experiments and calculations," 182.

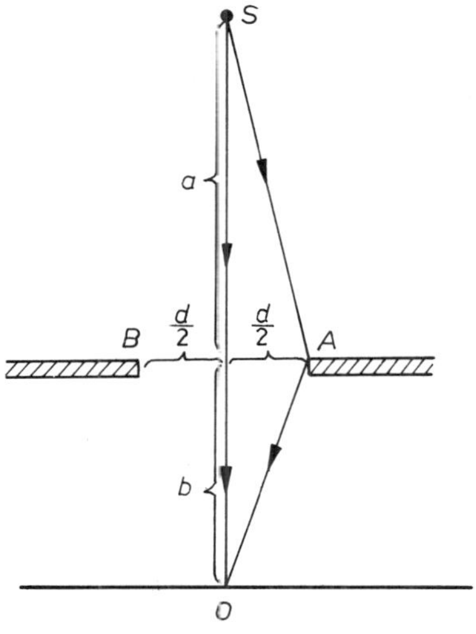

Fig. 27
An illustration of Young's theory of Newton's diffraction experiment with a slit

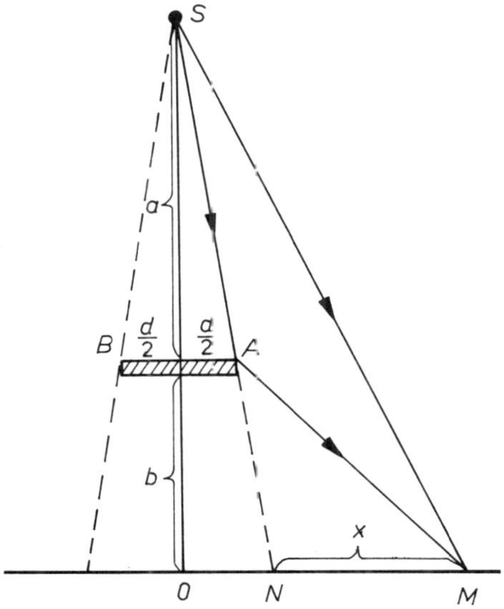

Fig. 28
An illustration of Young's theory of diffraction (the external fringes)

shadow. Young's "interval" is one half of the wavelength. Now, the positions of the bright and dark fringes are

$$x_{\max} = \sqrt{\frac{2\,(m-1)\,\lambda b\,(a+b)}{a}}; \quad m = 1, 2, 3,\ldots \tag{5.26}$$

$$x_{\min} = \sqrt{\frac{(2\,m-1)\,\lambda b\,(a+b)}{a}}; \quad m = 1, 2, 3,\ldots \tag{5.27}$$

When the positions are known, the wavelengths can be determined by two methods

$$\lambda = \frac{x^2_{\max} a}{2mb\,(a+b)}; \quad m = 1, 2, 3,\ldots \tag{5.28}$$

$$\lambda = \frac{x^2_{\min} a}{(2m-1)\,b\,(a+b)}; \quad m = 1, 2, 3,\ldots \tag{5.29}$$

The wavelengths calculated in this way agree very well with Young's results (see Table 4, observation 3). We can thus conclude that he had similar formulae.

Let us now analyze Young's results. Since *all* the wavelengths are shifted to the red end of the spectrum, we may suspect a flaw in the theory. I will show that Young failed because he neglected the phase change for the reflected ray. Indeed, after introducing this hypothesis, we obtain instead of eqs. (5.26–5.27)

$$x_{\max} = \sqrt{\frac{(2\,m-1)\,\lambda b\,(a+b)}{a}};; \quad m = 1, 2, 3,\ldots \tag{5.30}$$

$$x_{\min} = \sqrt{\frac{2m\lambda b\,(a+b)}{a}}; \quad m = 0, 1, 2,\ldots \tag{5.31}$$

Thus, the maxima and minima exchange places relative to eqs.(5.26–5.27), and at the border of the shadows we have a dark fringe *(m=0)* in eq. (5.31). The next three equations for the wavelength are the modified eqs. (5.30), (5.31) and (5.24), respectively

$$\lambda = \frac{x^2_{\max} a}{(2m-1)\,b\,(a+b)}; \quad m = 1, 2, 3,\ldots \tag{5.32}$$

$$\lambda = \frac{x^2_{\max} a}{2mb\,(a+b)}; \quad m = 1, 2, 3,\ldots \tag{5.33}$$

$$\lambda = \frac{d^2\,(a+b)}{8ab}. \tag{5.34}$$

The wavelengths calculated according to eqs.(5.32–5.34) are presented in the last column of the table 4. They are much better than Young's results: the wavelengths are closer to those derived from the colors of thin films, and their difference is much smaller.

Young did not consider the discrepancy in his results to be too large; in his view, he proved that the wavelength calculated from different observations is "either accurately or very nearly constant."[78] After comparing these wavelengths with those derived from observations of thin films, he announced that in spite of their differing by one-eighth, "this appears to be a coincidence fully sufficient to authorise us to attribute these two classes of phenomena to the same cause."[79] This statement has been used by some historians to accuse Young in an amateurish way of doing physical research. In fact, Young displayed here a rare perceptiveness. He suspected that something was wrong with his theory, which particularly affected the fringes of the first few orders. However, he was convinced that the fault was not with the principle of interference but with some supplementary hypothesis; thus he decided to publish his results in a hope that other researchers will improve them. Actually, he did it himself.

In 1813 he publicly accepted the hypothesis of phase inversion to reconcile his theory with observations.[80] He realized then that what he had considered, following Grimaldi and Newton, as an augmentation of a shadow by diffraction, actually was the first external dark fringe. There is, however, evidence that he did recalculate all the wavelenghts, using the phase change, long before 1813. He said in the *Lectures* that,

> Upon this theory it follows that the *distance of the first dark fringe from the shadow be half as great as that of the fourth*, the difference of the lengths of the different paths of the light being as the squares of those distances; and the experiment precisely confirms this calculation, with the same slight correction only as is required in all other cases; the distances of the first fringes being always a little increased.[81]

---

[78] Ibid., 183.
[79] Ibid.
[80] Young, "On blood and pus," 355.
[81] Young, *Lectures* 1:467, italics added.
I first described Young's error in N. Kipnis, "The first theory of diffraction by Young and Fresnel" (in Russian) in *Akademiia Nauk SSSR. Institut Istorii Estestvoznaniia i Tekhniki. Trudy XVII nauchnoi konferentsii. Moskva. 1974. Sektsiia istorii fiziki* (Moscow: Viniti, 1975), 92–95, no.2456–75 Dep. At the time I was unaware that Young corrected the error.

Indeed, it follows from eq. (5.31) that

$$\frac{\chi_{\min 1}}{\chi_{\min 4}} = \frac{\sqrt{2}}{\sqrt{8}} = \frac{1}{2}, \tag{5.35}$$

and all fringes, except for the first one, yield wavelengths very close to the actual one (within Young's measure of precision). Unfortunately, Young never published his revised calculations of the wavelength.

Thus, by 1807 Young apparently had a satisfactory explanation of the external fringes. Neither his initial omission of the phase change nor his subsequent correction of it have been previously recognized by historians, who have supposed that Young had introduced this hypothesis right from the start just as Fresnel did. It must be noted that Young's theory of diffraction was more mathematical than physical: to determine the positions of fringes produced by interference, he needed only two points which played the role of centers of spherical waves. He found that for the internal fringes these centers were at the opposite edges of the body, while for the external fringes they coincided with an edge and the luminous other than that source. He could not present any physical reason for this choice other than that the hypothesis worked. Sometimes, this approach led Young to physically impossible consequences. For instance, to produce fringes within the shadow of a body, Young's inflected rays must pass through the body itself.

V.8 The two-slit experiment

Young described in his *Lectures* an observation which he considered to be the simplest case of interference of light:

> . . . a beam of homogeneous light falls on a screen in which there are two very small holes or slits, which may be considered as centres of divergence, from whence the light is diffracted in every direction. In this case, when the two newly formed beams are received on a surface placed so as to intercept them, their light is divided by dark stripes into portions nearly equal, but becoming wider as the surface is more remote from the apertures, so as to subtend very nearly equal angles from the apertures at all distances, and wider also in the same proportion as the apertures are closer to each other. The middle of the two portions is always light, and the bright stripes on each side are at such distances, that the light,

> coming to them from one of the apertures, must have passed through a longer space than that which comes from the other, by an interval which is equal to the breadth of one, two, three, or more of the supposed undulations, while the intervening dark spaces correspond to a difference of half a supposed undulation, of one and a half, of two and a half, or more.[82]

This classic experiment has traditionally been presented in physics textbooks as having played an important role in the establishment of the wave theory of light. Recently, however, Worrall has introduced a series of arguments to show that Young never performed the two-slit experiment: 1) Young never explicity claimed to have done it; 2) he gave no numerical details; 3) he provided no details about the source of light and other conditions necessary to make this experiment succeed; and 4) he mentioned two-slit interference only in his *Lectures* and never again.[83]

Although Worrall's arguments are valid, they do not necessarily lead to his conclusion and may be explained differently. The lack of numerical and experimental detail is typical for experiments described in the *Lectures*.[84] Young tended to simplify his *Lectures* as much as possible, and for this purpose he eliminated most mathematical and technical details. He provided those who were interested in such details with a mathematical Supplement and a large bibliography. Therefore, the descriptions of experiments in the *Lectures* necessarily differed from those in his papers in the *Philosophical Transactions*. In the next chapter I will answer Worrall's other two points. I support his contention that the passage cited above describes a thought experiment rather than an actual one. In my view, it would have been difficult for Young to prove experimentally that the breadth of a fringe δ is directly proportional to the distance b between the slits and the observation screen and inversely proportional to the interval d between the slits (see my Fig. 29). It is more likely that he referred to the following equation, derived theoretically:

$$\delta = \lambda \frac{b}{d}, \qquad (5.36)$$

[82] Ibid., 464.

[83] Worral, "Thomas Young," 152–56.

[84] Cantor has made this point in his "Historiography of 'Georgian optics'," *History of Science* 16 (1978): 12, n. 19.

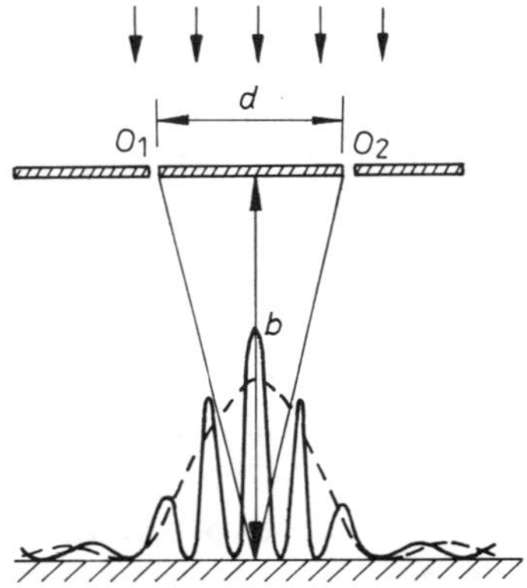

Fig. 29
The distribution of maxima and minima in the two-slit experiment: the dotted curve represents the diffraction fringes while the solid line represents the interference fringes

where λ is the wavelength.[85] However, granting that this passage is nothing more than a thought experiment, it does not mean that Young never actually performed the two-slit experiment.

There is evidence that he, in fact, did make this observation. An analysis of several paragraphs which follow his decription of two-slit interference in Lecture XXXIX, "On the nature of light," reveals that Young was attempting to found an explanation for a number of diffraction phenomena on his explanation of two-slit interference.[86] According to Young, this experiment is the simplest case of the interference of two rays, since two small holes can be considered to be two centers of waves. Two slits may be treated as a number of pairs of luminous centers which produce the same interference pattern as a single pair, but of a greater intensity. In the following paragraphs Young shows how different cases of diffraction may be reduced to the interference of light from two linear sources. Young argues, for instance, that by removing a part of the screen on the outside of the slits, a "narrow body" is obtained whose edges coincide with the former

[85] See, for instance, D. Halliday, R. Resnick, *Physics* (New York: J. Wiley & Sons, 1978), 998.
[86] Young, *Lectures* 1:465–67.

slits. In this way diffraction by a narrow body can be treated as the result of interference of rays coming from the opposite edges of the body.

To simplify references I will number all the paragraphs in Young's Lecture XXXIX beginning with the "two-slit thought experiment" and give a brief description of them. Young's treatment of all phenomena is both theoretical and experimental, though the details of observations are not provided. The subject of the second and third paragraphs is not stated explicitly, and I will return to them shortly. The fourth paragraph deals with the internal diffraction fringes produced by a narrow body; the fifth describes diffraction by a single parallel slit; the sixth is devoted to diffraction by a striated surface; the seventh treats diffraction by a corner of a body; the eighth examines the external diffraction fringes; and the next paragraphs deal wih the colors of thin plates, natural bodies, mixed plates, supernumerary rainbows, and thick and double plates. It is clear from this list that the two-slit experiment is missing, and, according to the logic of the exposition, this experiment should be the subject of the second paragraph (or of both the second and the third). I will now present the evidence to support my conjecture.

The second paragraph states that,

> The combination of two portions of white or mixed light, when viewed at a great distance, exhibits a few white and black stripes, corresponding to this interval; although, upon closer inspection, the distinct effect of an infinite number of stripes of different breadths appear to be compounded together, so as to produce a beautiful diversity of tints, passing by degrees into each other. The central whiteness is first changed to a yellowish, and then to a tawny colour, succeeded by crimson, and by violet and blue, which together appear, when seen at a distance, as a dark stripe; after this a green light appears, and the dark space beyond it has crimson hue; the subsequent lights are all more or less green, the dark spaces purple and reddish; and the red light appears so far to predominate in all these effects, that the red or purple stripes occupy nearly the same place in the mixed fringes as if their light were received separately.[87]

This description is too detailed for a thought experiment, and undoubtedly Young saw these colors. However, it is unclear whether he used two slits or just one, because in both cases the colors (the "diffracton fringes") would have been the same. The only unique feature of the two-slit phenomenon is a number of

[87] Ibid., 465

equidistant narrow white and black "interference fringes" seen in the center of the diffraction pattern. Since Young did not mention them, it would appear that he was working with a single slit. However, his third paragraph refutes this idea:

> The comparison of the results of this theory with experiment fully establishes their general coincidence; it indicates, however, a slight correction in some of the measures, on account on some unknown cause, perhaps connected with the intimate nature of diffraction, which uniformly occasions the portions of light, proceeding in a direction very nearly rectilinear, to be divided into stripes, formed by the light which is more bent (Plate XXX, Fig. 442, 443).[88]

There are two reasons to believe that Young is here speaking of the two-slit phenomenon. First, Fig. 442, to which he refers (see Fig. 30), shows two slits. Second, Young did not have a theory for diffraction by a single slit. According to his idea that external fringes are produced by the interference of direct and reflected light, in this case he had to consider *three* rays, since both edges reflect light. In

Fig. 30
Young's diagram of his two-slit experiment displays the fringes as well as their loci (from *Lectures* 1: Fig. 442)

[88] Ibid.

the particular case of a central fringe the paths of the two reflected rays are the same, and thus the problem can be reduced to two-ray interference. This is the only case of diffraction by a slit that Young investigated. Therefore, the "theory" mentioned by Young could have been only the theory of interference from two luminous centers, expressed in eq. (5.36). According to this theory, Young expected to find equidistant fringes, while the experiment showed the central fringe to be "a litte dilated" as compared to others: that is why Young decided that the theory is valid and needs only a "slight correction."[89]

Thus we have an apparent contradiction, for if the first and third paragraphs refer to two slits, the second cannot deal with a single slit. This paradox can be resolved if we assume that Young was working with two slits but for some reason observed only a part of the complete interference pattern, namely the diffraction fringes, while missing the interference fringes. Details on the colors observed show that the fringes were vivid and quite broad. The easiest way to see such fringes is by looking *through* the slits, held near the eye, at a distant source of light (a candle flame, for instance, as in Fig. 31).[90] To obtain bright fringes on a

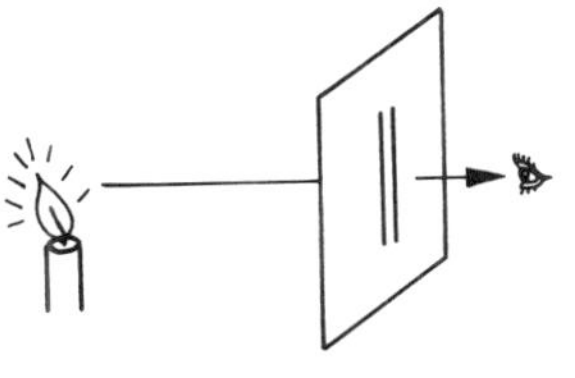

Fig. 31
Direct method of observing fringes in the two-slit experiment

[89] Ibid., plate XXX, Fig. 442 and the caption to it on p. 787. The second paragraph emphasizes a predominance of reddish colors: such coloration is characteristic for diffraction fringes produced by each slit separately; the interference fringes obtained in white light display more balance between reddish and bluish colors.

[90] I used a candle flame not only as a source of light but also to make a double slit. After covering a microscopic slide with soot, I drew with a razor blade two lines on the smoked surface, either parallel or intersecting at a very small angle. Such double slits were tested at the Bakken Museum by high school teachers and found easy to make and excellent for qualitative observations. With a V-slit, described above, one simply cannot miss the interference fringes. This technique was available to Young, who also was familiar with the "direct" method of observing diffraction produced by a hair, or thin fibers, or small particles.

white screen using sunlight he had to have a larger opening in the shutter than a pinhole. Whatever method Young utilized, it appears that he missed the interference fringes because his slits were too far from one another.[91]

Thus Young did experiment with two slits, and he used both white and monochromatic light.[92] However he did not discover the interference fringes: he confused them with diffraction fringes. Because the interval between the slits was too large, he could see only diffraction fringes produced by each slit separately. One of the obvious reasons for his mistake was the qualitative character of his observation: if he had measured the distance between the observed fringes, he would have immediately realized that they were of the wrong kind.

Section III: Young on coherence of light

Ernst Mach (1838–1916) believed that "Young evolved a practically complete theory of the conditions for interference," while, in Worrall's view, "Young had no clear ideas on what was later called coherence."[93] Since both opinions are based on a few selected passages from Young, the subject requires a more thorough investigation.

As mentioned in I.4, coherent wave must have: 1) equal frequency; 2) common origin; 3) small angle between them; 4) small path difference; and 5) very small light source (in some cases of diffraction). The first three conditions of coherence can be found in Young's formulations of the principle of interference Y(1)

[91] For instance, by looking through a glass with two 0.1 mm slits at a candle flame, I saw the interference fringes when the slits were 0.75 mm apart but did not see them when the distance between the slits was 1.5 mm.

[92] Young twice mentioned monochromatic light in relation to two-slit interference and showed green maxima in his Fig. 442 in the original edition of his *Lectures*. It is not clear how he monochromatized light, for he applied a prism only to observe the whole spectrum (see *Lectures* 1: Plate XXX, Fig. 443). It is not improbable, however, that he referred to monochromatic light only to simplify his explanation of the theory of two-slit interference.

[93] See Mach, *Principles of Optics*, 153 (to Mach, "practically complete theory" meant the condition of direction and the condition of common origin); Worrall, 154.
Young did not use any special term to designate the conditions that light must fulfill to produce an interference phenomenon. It seems that Felix Billet was the first to use the terms "cohérents rayons" and "incohérence"; see his *Traité d'optique physique* 2 vols. (Paris, 1858), 1: 63, 444–5. Paul Drude applied "Kohärenz"; see his *Lehrbuch der Optik* (Leipzig, 1900), and also *The Theory of Optics*, trans. from the German (New York, 1907), 134.

through Y(6) while the last two are missing. Does it mean that Young was unaware of the two? To answer this question we have to know how reliable the definitions of the principle of interference are as a source of information about coherence. For instance, the condition of a common origin appeared only in Y(5), which may mean that prior to 1807 Young accepted a possibility of interference from two sources. However, this condition disappeared from Y(6). Does it mean that Young changed his mind again and returned to interference from two sources? Another example: the condition of frequency appeared only in Y(1). To bypass this obstacle historians supposed that the term "same light" in Y(2) and Y(4) means "same frequency." This approach, however, does not solve the problem of Y(3) (is "same pencil" identical with "same light"?), and even less so of Y(6), where there is neither "same light" nor "same pencil." It seems that identifying Young's views on coherence from the formulations Y(1) through Y(6) is a hopeless business.

Fortunately, there is another, much less controversial, type of evidence of Young's *knowledge* of the conditions of coherence. I prefer "knowledge" to "understanding," the term favored by some historians. The latter means a "correct" explanation of the condition of coherence and implies that without it the conditions could not be applied. In fact, although Young explained neither the conditions of coherence themselves nor how he obtained them, there are *proofs* that he used them.

The simplest and the best proof that Young knew several of them is the success of his theory. If one accepts the conclusion of Ch. IV that Young gave a satisfactory for his time quantitative account of a number of optical phenomena, one has no choice but to admit that Young knew several conditions of coherence. The reason for this is that no theory of interference can be formulated without these conditions.

Indeed, the theory provides us with locations of colored fringes which must be tested experimentally. We obtain them from the equation for the path difference of interfering rays, which we derive by taking into consideration a number of hypotheses mentioned in IV.5, including some conditions of coherence. For instance, we use interfering rays of the same frequency, which come from the same point. Mathematically, this is easy to justify: if we do otherwise, the problem becomes much more complex if not indeterminate. Indeed, if we use different frequencies, how to select them? If each ray of light has a separate origin, how to select the proper pair? Mathematics, however, is indifferent to the length of the path difference or to the size of a luminous source. The limitations imposed on these parameters can be found by comparing the predicted positions of fringes with observations. If, for instance, the actual angular distance of two colors differs from the calculated one, we could have selected a wrong pair of rays.

From the agreement of Young's equations reconstructed in Ch. V with experiment, we may infer that he used the condition of frequency and those of a common origin, and of path difference.

Young's diagrams in his "Theory of light" indicate the interfering rays as coming from the same point, either at a finite or at an infinite distance from the observer.[94] A "luminous *point*" is shown even in the case of thin films, where the actual light source was an extended body. This helps to explain the meaning of the condition of common origin. The point source in the diagrams refers to a *theoretical assumption* rather than to a small size of a real luminous body: whatever the size of the source of light, we may replace it with a point taken arbitrarily on its surface.

I will now discuss in some detail how the conditions of coherence appear in Young's papers otherwise than in equations and diagrams, including those unaccounted before, and I will also speculate on how Young could explain them.

V.9 The condition of frequency

Although Young mentioned explicitly the condition of frequency only in Y(1) (see IV.5), there are indications that he implied it in his other formulations of the principle of interference. For instance, the "appearance and disappearance of various colours" in Y(2) and Y(4) means that he deals there with two beams of white light. Thus, "same light" may mean there, for instance, red rays of one white beam interfering with red rays of the other beam, and so on. However, even if "same light" refers not to the same color but to the same origin, we have another proof that all his formulations contain the condition of frequency. Indeed, Young's point that each color appears where the path difference is a multiple of the wavelength specific for this color makes sense only if the wavelength is the *same* for both interfering waves.

Young never explained why the interfering light waves must have the same frequency. Certainly, this condition cannot be obtained experimentally, for any real beam contains a number of different wavelengths, and it is impossible to isolate two beams of light of *one* wavelength.[95] The meaning of the condition of frequency is that when two beams of white light meet, we must *consider* the interference between red rays of the first beam and red rays of the second, green rays

[94] Young, "Theory of light," 169, Fig. 129, 130

[95] Mach probably excluded this condition as empirically unverifiable; *Principles of Optics*, 153.

of the first and green of the second, and so on, but not between rays of different colors.

Young did not mention how he recognized the impossibility of interference of light of different colors. He might have tried to extend his explanation of beats of sound to light of different frequencies and failed.[96] Whatever his exact arguments, in 1801 Young postulated that interference of light of different colors is impossible. Young needed the condition of frequency to account for the production of *colors*. According to him, interference of light is a means for separating the colors contained in white light (like the dispersion of light in a prism).[97] His idea is that when two beams of white light meet at any point, the condition of the destructive interference is fulfilled for some of the colors, and these colors disappear. As a result, a white luminous spot becomes colored. Unfortunately, Young did not explain this process clearly, and some of his contemporaries understood him to be contradicting Newton's theory of colors, for Newton demonstrated that a mixture of colors produces white light, while Young apparently claimed the opposite, that is, that a mixture of two beams of white light formed colors. I will return to this point in the next chapter.

V.10 The condition of direction

The requirement that interfering rays must have almost the same direction first appeared in the "Theory of light."[98] Young considered this condition to be very important, since he repeated it in all formulations of the principle of interference. It is unlikely that he justified the optical condition of direction in the same way as its mechanical counterpart, namely, that an increase of the angle between the vectors representing two vibrations reduces the amplitude of compound vibrations until it becomes imperceptible, because this effect could account for the disappearance of the fringes only at a very large angle between the interfering

[96] By using Young's table of the frequencies of light ("Theory of light") one can calculate the "combination tones" of light for interference of, say, red and violet light and find that it can be infrared, or ultraviolet which cannot be seen.

[97] Young, "Theory of light," 148–49.

[98] Young, "Theory of light," 163.
Worrall believes that the requirement of almost parallel rays proves Young's inability to explain the direction of the resulting vibrations in the case of a large angle between the rays; "Thomas Young," 138–39. This case would have been, however, of no interest to Young and Fresnel, for, according to eq. (5.37), the larger the angle the more narrow are the fringes until they become invisible, which means that such rays do not interfere. To Worrall, however, the invisibility of fringes does not mean absence of interference.

rays. Young could have developed the same explanation of the condition of direction as Fresnel did, namely, that an increase of the angle between the rays causes a decrease in the fringe's breadth, and the fringes can become too narrow to be seen. Mathematically, Fresnel's condition may be presented as

$$\delta = \frac{\lambda}{\alpha} \tag{5.37}$$

where δ is the fringes' breadth, λ is the wavelength, and α is the angle between the interfering rays. Young could have derived eq. (5.37) from eq. (5.36) for two-slit interference by substituting $\alpha = d/b$ (see Fig. 29), where d is the distance between two point sources, and b is the distance from the sources to the observation screen.

V.11 The condition of path difference

Between May and July 1801, while attempting to apply his principle of interference to the colors of thick plates, Young realized that the path difference of interfering rays cannot be large. However, neither in the *Syllabus* nor in the "Theory of light" did he formulate this requirement directly. It appears in his objections to Newton in the "Theory of light." When applying his theory of fits Newton could not answer the question why the colored rings vanish when a thin film's thickness exceeds 1/15 000 of an inch, whereas they appear in a plate 1/4 inch thick. Actually, he answered the first part of it: the disappearance of the fringes in thin films is due to an overlapping of the fringes of different orders; the thicker the plate, the greater the mixture of different colors, and finally there are no colors left. According to Young, the production of colors of thick plates "is by no means, as Newton imagined, identical with the production of those of thin plates."[99] Young clarified this point in his *Lectures* by saying that

> the effect of a plate of any considerable thickness must be absolutely lost in white light, after ten or twelve alternations of colours at most.[100]

He means that in thick plates the rays regularly reflected at the two surfaces cannot interfere owing to a complete overlapping of colors of different orders. Young's refusal to treat inteference in thick plates analogously to that in thin

[99] Young, "Theory of light," 163.
[100] Young, *Lectures* 1:471.

films demonstrates his awareness of the necessity of small path difference. Young, however, was not correct in objecting to Newton's claim of an analogy between the colors of thin and thick plates. Newton considered them to be analogous because he could explain them with the same constant of periodicity for the same color, while Young focused on different procedures to calculate the radii of rings produced in the two phenomena.

Another strong proof of Young's use of the condition of path difference is his explanation of colors of "double plates." In 1799, Nicholson described periodical colors produced by two parallel plates of about the same thickness; in 1815, Brewster discovered similar colors created by identical thick plates making a very small angle between them, and Knox observed alternating colored fringes made by several lenses placed on one another.[101] Young explained that all these colors were the result of either a small difference in thickness of parallel plates, or a slightly different incidence of light on identical but not exactly parallel plates.[102] In all these cases, he selected only such interfering rays which have a very small path difference.

The fact that Young never mentioned the condition of a small path difference suggests that he considered this condition to be a consequence of another condition, possibly of the condition of direction. He could have derived from his explanation of the colors of thin films that, given the positions of the plate and of the observer, the angle between the interfering rays is proportional to the path difference. On the contrary, modern textbooks tend to view the condition of path difference as the primary one. Whichever the primary, we should remember that these two conditions are equivalent.

[101] Nicholson, "Experiments and remarks on certain ranges of colours hitherto unobserved . . .," *Nicholson's Journal* 2 (1799): 312–5; Brewster, "On a new species of coloured fringes produced by the reflection of light between two plates of glass of equal thickness," *Transactions of the Royal Society of Edinburgh* 7 (1815): 435–44; and Know, "On some phenomena of colours exhibited by thin plates," *Phil Trans*., 1815: 161–81.

[102] In 1807, Young wrote (*Lectures* 2: 317) that Nicholson's colors arise,

> probably from a slight differnce in the thickness of the glasses, the rays twice reflected within the first glass only, interfering with the rays twice reflected in the second only. The analogy with the colours of thin plates is wholly foreign to the subject.

In 1815, he gave a brief explanation of the colors discovered by Brewster and Knox (Young to Brewster, September 13, 1815: *Misc. Works* in his *Chromatics*, 292–93.

V.12 The condition of common origin

In all cases of interference examined above Young considered interfering rays to originate from the same point. Some historians suggest, however, that he sometimes utilized interference of light from two sources. They have brought forth three arguments to support this claim. First, that Proposition VIII in the "Theory of light" mentioned two sources. Secondly, that Young introduced a *necessity* of a common origin only in 1807 (see Y(5)). Finally, that he admitted that two candle flames can create an interference pattern.

As to Proposition VIII, I have already shown that it refers to waves in general, rather than to light, and for some waves, such as sound or water waves, two sources can be used indeed. The second point is actually irrelevant unless one can prove that prior to 1807 Young accepted the possibility of interference of light from different sources. Since this is also the essence of the third argument, let us examine it carefully.

In October 1804, Brougham noted:

> If his [Young's] theory is worth a thought from its author, . . . it must follow, that we can always form colored fringes, by causing two beams of white light to interfere. In other words, by *doubling the quantity of light* on any place, we can cover it with colored fringes; or which is the same thing, the colored fringes are nothing absolute, but a mere relative idea, like size and intensity, denoting the increase of any given density of illumination.[103]

Two months later Young replied:

> The reviewer has cursorily observed that if the *law* [of interference] were true, every surface opposed to the light of two candles would appear to be covered with fringes of colours. Let us suppose the assertion true – what will be the consequence? In all common cases the fringes will demonstrably be *invisible*; since, if we calculate the length and breadth of each fringe, we shall find that a hundred such fringes would not cover the point of a needle.[104]

[103] [Brougham], "Dr. Young's Bakerian Lecture," *Edinburgh Review* 5 (1804): 99, italics added.
[104] Young, "Reply," 203, italics added.

What Young is saying is that even if two candle flames produce the fringes, they would be too narrow to be seen. Apparently, he treats two candle flames as two coherent sources of light, which is not the case. For this reason, Worrall accuses Young of misunderstanding coherence.[105]

In fact, Young's error lies not in assuming the interference of light from two candle flames (he says they do *not* produce fringes, which, means, in our terms, that light does not interfere) but in his explanation of the reason for this non-interference. Worrall notes that the correct explanation was given by Fresnel who supposed that light waves experience frequent random changes of their phase, the result of which is a division of an infinite wave into groups of waves (later called "wave trains") of finite length. Since different luminous points emit these wave trains non-synchronously, their phase difference at any point will change so rapidly that the eye perceives only an average illumination.[106]

Incidentally, there was no need for Young to put himself in this embarassing position, for Brougham did not ask about the *origin* of the "two beams" (it will be shown further that he was not concerned with coherence). That Young misread Brougham means that he had already been preoccupied with this problem. The character of Brougham's criticism made it impossible for Young to acknowledge his ignorance, and to save his face he offered an explanation which at least had some sense. Later be found that his answer was poor and decided to *postulate* in Y(5) that interfering rays must come out of the same origin.

This does not necessarily mean that by the time he published the *Lectures* Young had not possessed a solution similar to Fresnel's. The key point of Fresnel's theory was an irregular interruption of emission of light. Young also played with the idea of a discontinuous emission. In his "Theory of light" he supposed that waves of different colors enter the eye at different moments.[107] In 1807, however, he expressed it more strongly:

> there is no reason to suppose the undulations of light continuous: their intermissions may easily be a million million times greater than the duration of each parcel of undulations.[108]

It is difficult to imagine that Young could have missed the fact that the continuous model of light contradicts the condition of path difference: two infinite

---

105 Worrall, 140. When I discovered this passage in 1974 I interpreted it similarly to Worrall, i.e. that Young did not grasp coherence, and so I reported in May 1974 to the All-Union conference of historians of exact sciences in Tambov, U.S.S.R. It takes time to understand Young.

106 See Fresnel's explanation in VII.1(e).

107 Young, "Theory of light," 148–9.

108 Young, *Lectures*, 2: 544.

waves of the same frequency would interfere even at a very large path difference. Thus, by 1807, Young probably possessed an explanation of the non-interference of light from different sources similar to Fresnel's, for after assuming a discontinuity of the emission of light one *has* to admit that emission from different points is not synchronous. That he did not publish this explanation in his *Lectures* or elsewhere does not mean much, since he never explained *any* condition of coherence. Moreover, he had a reason for not publishing it: the hypothesis of discontinuous light had then no experimental support whatsoever, and Young had already suffered from accusations in an excessive hypothesizing.

One may note that an admission of discontinuity of light is incompatible with Young's explanation of the condition of path difference by overlapping of waves of different frequency. Indeed, in this case waves must be infinite, however Fresnel faced the same problem and found nothing wrong in using both models of light (see VII.8).

By all probability, in 1801, Young was not positive about the discontinuity of light, and this raises the question: how could he discover the necessity of a common origin without explaining it? We already know that a common origin is required by mathematics. Besides, Young could have conceived simple and convincing physical arguments against a possibility of interference of rays coming from different points.

He probably never tried to place an opaque screen with two small holes before a source of light. At least, he did not do it with sunlight (see V.13). Generally, such an experiment would be inconclusive anyway, because the holes may become coherent sources. Perhaps, Young thought that he did not need any new experiments because the known ones were sufficient to refute the assumption that interfering rays may originate from different points.

First, changing the distance between the source of light and the screen where fringes were observed (for instance, two glasses pressed together) would change the path difference and therefore the interference pattern. This contradicts observations. Secondly, since each pair of luminous points produces its own set of fringes, an overlap of many such sets will obliterate most colors. One may expect this effect to be lesser for a smaller luminous body. In reality, reducing the size of a luminous source does not improve the visibility of Newton's rings. Finally, a small luminous body (for instance, a pinhole placed before a candle flame) when viewed at a distance should display colors, at least at the center and at the rim, which would change at a slightest movement of the head. Experiments do not confirm this either.

Now, after establishing that Young correctly applied the condition of common origin since 1801, we can try to determine where this condition is "hidden" in different formulations of the principle of interference. For instance, the expression "two portions of the same pencil [Young's "pencil" means a beam of light]" in Y(3) may contain it, for it implies that a single beam of light is split into two parts which interfere. Another expression "two . . . portions of light . . . have been separated and coincide again" from Y(6) has the same meaning and seems to be even closer to the modern view: a wave is split in two parts, which combine after traveling unequal distances. Young's insistence on interfering rays passing *different* routes also points out to their common origin, for if the rays come from two points, the difference of their routes is secured automatically.

We may now assert that the differences in Young's formulations do not necessarily reflect changing views. Sometimes, he uses different terms to describe the same idea. For instance, the expressions "same light" in Y(2) and "same pencil" in Y(3) belong to the same period (late 1801) and thus should have the same meaning. Another example: when Young reprinted his optical articles in his *Lectures* with some revisions, none of the revisions affected the principle of interference.[109] As a result, we see several apparently inconsistent formulations Y(1) through Y(5) published at the same time! Certainly, Young did not think they contradicted one another, which means that he changed the words but not the ideas. It seems Young simply was not concerned with rigorous formulations, and this lack of care should not be the reason for attributing to him the views which he never held.

V.13 The condition of the size of a light source

Let us now see what kinds of problems Young encountered when he substituted a luminous point for a real light source. Diffraction evidently does not create any difficulty, since a very small secondary source employed in observations of this sort (a very small aperture, or the focus of a lens collecting sunlight) may be considered a good approximation to a point. The situation is not so simple with observations of the colors of thin films where a broad source of light (the blue sky or white clouds) is usually applied. Fortunately, in this case all light rays coming to the plate from the *same direction* contribute to the formation of the same fringe. Therefore, in order to determine the location of this fringe, the source may be replaced by a single point. However, the problem becomes more difficult when it is necessary to take into consideration the rays traveling from different points of the source in different directions. In this case, according to the modern ap-

[109] Young, *Lectures*, 2: 531–54, 613–31, 633–38, 639–48.

proach, each point of the source is considered to produce its own system of fringes. An overlap of different systems reduces the visibility of fringes, and the bigger the angular size of the source the greater the effect. Let us see whether Young was aware of such an approach.

As mentioned above, in all his early papers Young applied the approximation of a point source even when he actually used the whole disc of the sun or a candle's flame. Since this approximation showed satisfactory agreement with observations, he simply ignored the dimensions of luminous bodies. He was able, though, if he needed it, to take these dimensions into consideration. This is seen in his discussion of the two-candle flame interference in his reply to Brougham, where he mentioned the "length" and "breadth" of the fringes. With two point sources the loci of the maxima are a pair of hyperboloids. In fact, we see a very small part of this surface, only where it intersects the retina. If the two sources are linear, different pairs of their points produce images at different parts of the retina. Finally, we obtain a fringe – the image of the source of light – whose length is proportional to the length of the source. Therefore, when speaking of the fringes' length, Young took into consideration the length of the flame. It would be natural to suppose that he also considered the breadth of the flame, but we have no evidence of it. However, even if Young considered the dimensions of two bodies when discussing interference of rays coming from *different* bodies, he did not do this when the luminous centers belonged to the *same* body, as shown in his discussion of an experiment by Grimaldi.

Grimaldi let sunlight enter a dark room through two small holes in a shutter and observed dark arcs where the two images of the sun intersected on a white screen (Fig. 32). He concluded from this that when one light is added to another light, darkness is sometimes produced.[110] This remark has led many authors to attribute to Grimaldi the first observation of interference in the two-aperture experiment. In fact, Grimaldi's diagram shows no trace of hyperbolical fringes, which must be present if the two holes were coherent sources of light. Émile Ver-

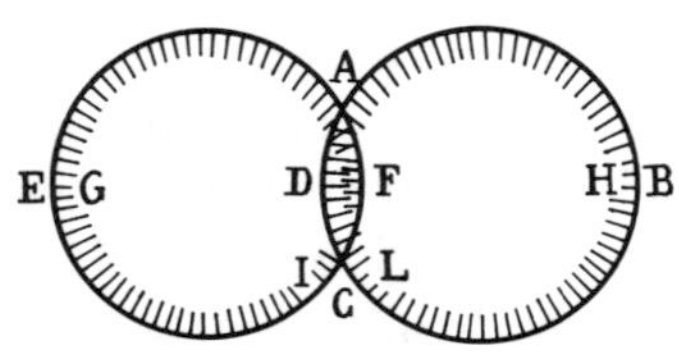

Fig. 32
Grimaldi's illustration of his two-hole experiment (from his *De lumine*)

[110] Grimaldi, *De lumine*, 187. See also this section translated in *Ann. de Chim.* 10 (1819): 306–12. Grimaldi emphasized that the two holes must be sufficiently distant from one another (p. 306).

det (1824–1866) calculated that the two holes in Grimaldi's observation could be treated as coherent sources only if they were distant less than 0.05 mm, which obviously was not the case.[111] He believed that Grimaldi's experiment did not involve interference and could be explained by the effect of contrast.[112] He demonstrated (see VIII.7) that two apertures at the distance d from one another would be coherent secondary sources if the angular diameter of the luminous body does not exceed

$$\varphi = \frac{\lambda}{d}, \tag{5.38}$$

where λ is the wavelength, and φ is the angular diameter. To make interference from two slits possible, Verdet concluded, Young had to reduce the angular size of the sun's disc by using a narrow diaphragm between the sun and slits or by collecting sunlight at the focus of a lens, which becomes the secondary source.[113] Following Verdet, many authors have also attributed the use of the additional diaphragm to Young. However, if Young applied direct method of observation (see V.8), he could do without a diaphragm and obtain fringes even when using such a luminous source as the whole solar disc.[114]

If, as I have suggested, he confused interference fringes with diffraction fringes, he could have concluded that the visibility of fringes does not depend on the angular size of the source, since diffraction fringes could be obtained with extended sources of light. Whatever the reason, Young believed in the independence of the visibility of fringes of the size of a luminous source. Indeed, in his Lecture XL "On the history of optics" Young says that,

> He [Grimaldi] had even *observed* that in some instances the light of one pencil tended to *extinguish* that of another, but he had not inquired in what cases and according to what laws such an *interference* must be expected.[115]

Thus Young did not see any essential difference between Grimaldi's experiment and his own, which proves that he did not realize the role of a light source's size. He evidently considered his two-slit experiment to be a simple modification of Grimaldi's with holes replaced by slits; and it is possible that Young borrowed

---

[111] Verdet, *Leçons* 1 : 106. Also see his "Introduction" to the *Oeuvres de Fresnel* 1 : XXI.

[112] Verdet, *Leçons* 1 : 52–53.

[113] Ibid.

[114] I observed interferences fringes by looking at the solar disc at sunrise through two slits spaced by 0.35 mm.

[115] Young, *Lectures* 1 : 476, italics added.

the very idea of his observation from Grimaldi, whose book he quoted in 1803. This could explain why Young presented the two-slit experiment in such a modest way and did not claim priority for it.

V.14 Summary

In 1801, Young already knew that the conditions of frequency, direction, path difference, and common origin are both necessary and sufficient. Although he knew that light from different origins cannot interfere, he could not explain it properly. Nor did he solve the problem of the size of a luminous body. Thus, Young possessed a considerable understanding of the concept of coherence, and except for Grimaldi's two-hole experiment, he always applied it correctly. Unfortunately, Young did not pass his knowledge of coherence on to his readers, since his presentation of this concept was neither lucid nor complete. Some of the conditions of coherence were stated repeatedly, while others were mentioned seldom or never. Since none of his works contained the complete set of conditions, a reader would have had to know all of Young's optical works. The lack of explanation could have produced various difficulties in grasping both theoretical and experimental problems. For instance, readers might have pondered in vain why rays from different points of the same body do not interfere. Young did not provide them with a method for determining whether a particular plate was sufficiently thin, a fiber sufficiently narrow, or two slits close enough to observe fringes; and this could have prevented them from repeating Young's observations and designing new experiments on interference of light.

Thus by 1801 Young's theory of interference was quite complete, for it contained all the principal hypotheses, including the phase inversion. For every new phenomenon, however, he had to find its specific pair of interfering rays. Most applications of his theory were quantitative, although in some cases the theory could not have been fully verified due to experimental difficulties. Young was the first after Newton to have undertaken a systematic quantitative study of periodical colors. He may be called one of the pioneers of the mathematization of physical optics in the nineteenth century.

Young was an inventive and careful experimenter, although he was concerned with an experiment only insofar as he needed it to demonstrate his theory. When he uncovered a significant discrepancy between the theory and phenomena, he worked to improve their agreement, mostly by changing some part of his theory, for instance, by introducing phase inversion. But he never doubted the principle of interference. His mathematical derivations were accurate, though his equations seem incomplete when compared to the modern ones. Apparently he emphasized only those parts of his equations which he considered easy to verify. All of Young's observations, except for the two-slit experiment, were correct. He considered a precision of about 10 to 12% to be satisfactory for observations of periodical colors and never tried to achieve a better agreement between his theory and experiment. He said that he had explained these colors

> in general perfectly *mathematical*, and always within the probable limits of the errors of observation.[116]

When compared to the *best* results of his contemporaries in different branches of physics this precision seems to be inadequate.[117] However, in the field of periodical colors Young was then the only scientist to compare his measurements to a theory, thus it is meaningless to call the agreement achieved by him insufficient. And, as shown in Ch.V, none of his contemporaries complained of a lack of precision in his observations.

Young's readers faced three major difficulties. First, to grasp new ideas they had to go through all of his works, since none of them presented a complete and detailed exposition of his theory. Secondly, they had to reconstruct the derivation of Young's equations. Thirdly, they had to guess the missing physical arguments. Consequently, they might have understood only those explanations that gave sufficient detail or were properly illustrated, and where the mathematical and physical reasonings were simple. These were, in my view, the colors of thin films, of striated surfaces, of halos, and external diffraction fringes. We will see in the next chapter which factors caused the greatest problems.

[116] Young, "Review of the *Mémoires of Arcueil* [1810]," *Misc. Works* 1:252, italics added.

[117] At the time there were not many physical laws verified by measurements. Coulomb's law was based on four measurements, with errors ranging from 0 to 10%. Henry Cavendish determined the density of the earth with an error of 4.6%. Wollaston's four measurements supporting Huygens's theory of double refraction differed by no more than 2%. I found only one example related to periodical colors: Brougham's results on diffraction differed from Newton's by 4%. Newton's results still were the best in the field of periodical colors. He claimed that his observations of the colors of thick plates and of the colors by diffraction differed from his theory by 1–3%. Young never disputed the exactness of Newton's measurements, but apparently he considered it exaggerated. Since Young did not have at this time any rivals in the measurements of fringes, there is no reason to call his results crude, as some historians do.

Chapter VI

The response to the principle of interference (1801–1815)

It has long been considered an historical anomaly that Young's wave theory was not accepted before 1816. Some scholars claim, however, that this response was not at all unusual, since the theory did not, in fact, deserve to be accepted.[1] As I have shown above, however, Young's theory was not so severely flawed that it *should* have been completely ignored or rejected. I intend to argue that the principle of interference was rejected at the time for two reasons: 1) some scientists considered it to be a part of the wave theory, which they disliked; 2) others agreed to separate the two but could not imagine a mechanical model of interference of light. It will be shown that both groups misunderstood the principle of interference as a non-Newtonian theory of colors.

My major sources for surveying the response to Young's theory will be British review journals, correspondence, and scientific works, although French and German response will also be briefly considered. I will divide my account into an early period (1801–1805), when all of Young's principal optical papers appeared as well as reviews of them; and a later period (1807–1815), during which the interest in Young's works increased. My analysis of the reaction to his work will focus on the following: 1) comprehension of his physical arguments and mathematical derivations, 2) evaluation of his experiments, and 3) comments on the style of his writing.

---

[1] See n.2, Ch.V.

VI.1 Early comments (1801–1805): general survey

The earliest known comments on the "Theory of light" came from Young's friend Wollaston. In his letter to Young of August 26, 1801, he says:

> I value your Bakerian much, but I cannot say that I have yet inserted the undulatory doctrine into my creed, and it may be some time before I repeat it with fluency.[2]

A few days later, in another letter, Wollaston specifies his impression of Young's theory:

> Young (Phenomenon) finds that these as well as many others facts yet [referring to colours of thin plates] unexplained become very intelligible upon the old hypothesis of etheral vibrations. I am inclined to think he will nearly prove that to be true doctrine . . .[3]

There are several points to note here. First, the "Theory of light" was finished no later than August. Secondly, the term "Bakerian" mentioned three months before the paper was read at the Royal Society implies that the Council charged Wollaston to examine it and decide whether the "Theory of light" deserved to be presented as a Bakerian Lecture. Thirdly, by singling out the explanation of the colors of thin films – the phenomenon which had been already explained – Wollaston, a quantitative physicist, shows his preference for Young's solution rather than Newton's.

Finally, it is important to note that Wollaston approved the paper and Young's effort to develop the wave theory despite his reservations about it. He was a staunch "emissionist" and was reluctant to discuss any speculations not supported by facts, as can be seen from his letter to Young: "I am certainly at all times very reluctant to utter & still more so to publish any mere conjectures."[4] Thus, Wollaston must have been much impressed with Young's theory to say:

> since the theory by which he [Huygens] was quided in his inquiries, affords (as has been lately shown by Dr. Young) a simple explanation of

---

2 Wollaston to Young, August 26, 1801: Royal Society, Young's correspondence, Ms 242. Peacock slightly changed this passage when quoting it (see *Misc. Works* 1:261n

3 Wollaston to Rev. Robert Hasted, August 30, 1801, lent by D. O. Wollaston (transcription): University College, London, L. F. Gilbert notes on Wollaston, Box 1, File 1, Encl. A, no. 18. The insert in square brackets is by transcriber.

4 Wollaston to Young, Monday, Ev. 17th: Royal Society, Young correspondence, Ms 242, no. 48. I date this letter as of November 17, 1800.

> *several phenomena not yet accounted for by any other hypothesis*, it must be admitted that it is entitled to a higher degree of consideration than it has in general received.[5]

This comment has a prehistory. Wollaston invented a new method for measuring refraction and applied it to many substances, including Iceland spar. He was not certain that his results for the extraordinary ray obtained for different directions in this crystal made any sense until Young recommended to him Huygens' theory of Iceland spar: Wollaston found a close agreement between this theory and his measurements.

Many reviews of Wollaston's paper on Iceland spar mentioned his support of the wave theory, and this was significant for Young's cause.[6] Although Young was very pleased with Wollaston's comment and frequently cited it, he did not exaggerate his rather limited approval of the wave theory. In fact, he doubted whether Wollaston fully appreciated his principle of interference and had actually adopted the wave theory.[7]

Another early response came from Humphey Davy (1778–1829), Young's associate at the Royal Institution (he was Professor of Chemistry). Davy wrote on November 14, 1801, two days after the reading of the "Theory of light" began, that,

> The Bakerian Lecture by Dr. Young, our Lecturer on Natural Philosophy, is now reading before the Royal Society. He attempts to revive the doctrine of Huygens and Euler, that light depends upon undulations of an ethereal medium. His proof (i.e. his presumptive proofs) are drawn from some *strong and curious analogies* he has discovered between light and sound.[8]

Davy himself was an "emissionist." In 1799, he presented a theory, in which light was a "matter of a peculiar kind" capable of moving with great velocity and entering into chemical combination with bodies.[9] Subsequently, he changed his mind and stuck to the "common" emission theory.[10]

---

[5] Wollaston, "Iceland crystal," 148, italics added. Read on June 24, 1802.

[6] See n.32 below.

[7] Young, "Reply," 213.

[8] Bence Jones, *The Royal Institution: its founder and its first professors* (London, 1871), 326, italics added.

[9] Davy, *Collected Works of Sir Humprey Davy*, 9 vols. (London, 1839), 2: 8, 10. Henceforth cited as *Coll. Works.*

[10] Davy, "A syllabus of a course of lectures delivered at the Royal Institution of Great Britain [1802]," *Coll. Works* 2:395.

In 1799, on the basis of his experiments on melting ice by friction, Davy suggested that heat was generated by a motion of the particles of bodies.[11] In 1802 he stated that vibrating particles of bodies produced undulations in the ether that filled all bodies and that these undulations propagated in the free ether in space and thus transmitted heat from the sun.[12] Young was interested in Davy's wave theory of heat, since he believed that light and heat are of the same nature, differing only in the frequency of their undulations. Although Davy did not share Young's wave theory of light, he found it necessary to mention in his "Syllabus" that this theory

> has been lately supported by some *important* arguments deduced by Dr. Young, from the analogy between the laws of known undulatory motions, and those of light.[13]

Having abandoned in 1802 all hopes for a rapid adoption of the wave theory, Young decided to change his strategy and to promote his principle of interference independently of a theory of light. On July 1, 1802, he presented a purely experimental article, "Production of colours," which brought forth a new support for the principle of interference. The opening paragraph of the paper stated that,

> Whatever opinion may be entertained of the theory of light and colours which I have lately had the honour of submitting to the Royal Society, it must at any rate be allowed that it has given birth to the discovery of a simple and general law, capable of explaining a number of the phenomena of coloured light, which without this law, would remain insulated and unitelligible.[14]

Although two more comments on the "Theory of light" appeared in the summer of 1802, one rather positive, and the other clearly negative, Young apparently was disappointed. He was hoping that his paper would stir up a public debate on the nature of light, but none of his reviewers wanted to go into it. His mood in the fall of 1802 is reflected in Davy's letter to Davies Gilbert of October 26, 1802:

---

[11] Davy, "Essay on heat, light, and the combinations of light, with a new theory of respiration [1799]," *Coll. Works* 2: 11–23.

[12] Davy, "Syllabus," 390–1.

[13] Davy, "Syllabus," 391, italics added.

[14] Young, "Production of colours," 170. Read on July 1, 1802.

> Have you seen the theory of my collegue, Dr. Young, on the undulations of an ethereal medium as the cause of light? It is not likely to be a popular hypothesis after what has been said by Newton concerning it. He would be very much flattered if you could offer *any observations upon it, whether for or against*.[15]

One more review came in December 1802, and it rejected Young's theory without examination. But finally, in January 1803, Young received the challenge that he wanted, though presented in a rather unexpected way. A review of the "Theory of light" and "Production of colours" in the *Edinburgh Review* set forth a detailed refutation of the wave theory. The reviewer (later correctly identified by Young as Henry Brougham) vigorously attacked Young, claiming that his "Theory of light" "contains nothing which deserves the name, either of experiment or discovery," because he is just playing with hypotheses.[16] According to Brougham, "the making of an hypothesis is not the discovery of a truth."[17] He also claimed that, "Dr. Young is by no means more successful in making observations and experiments than in forming systems."[18]

Young answered with the article "Experiments and calculations," read on November 24, 1803. He continued his new tactic of avoiding all hypotheses and presenting the principle of interference as a phenomenological rule:

> From the experiments and calculations which have been premised, we may be allowed to infer, that homogeneous light, at certain equal distances in the direction of its motion, is possessed of *opposite qualities*, capable of neutralising or destroying each other, and of extinguishing the light, where they happen to be united; that these qualities succeed each other alternately in successive concentric superficies, at distances which are constant for the same light, passing through the same medium.[19]

Young believed that with his new paper the principle of interference had been set on solid ground, and he implicitly invited Brougham to find a weak link in the chain of his arguments.[20] Brougham replied in October 1804 by dismissing Young's paper as "the feeble lucubrations . . . in which we have searched, with-

---

15 Jones, *Royal Institution*, 330. Italics added.
16 [Brougham], "Bakerian lecture on light and colours," *The Edinburgh Review* 1 (1803): 450.
17 Ibid., 452.
18 [Brougham], "Young on colours," 457.
19 Young, "Experiments and calculations," 187.
20 Ibid., 188.

out success, for some traces of learning, acuteness, and ingenuity"[21] Young's turn to a phenomenologic approach did not impress Brougham, who considered Young's "opposite qualities" to be "metaphysical absurdities."[22] From this review Young realized that any attempt to continue a scientific dispute with Brougham would have been meaningless, and two months later he published a separate pamphlet, his "Reply," in which he tried to justify himself before his friends and general public. The last reviews of the "Experiments and calculations" appeared shortly afterwards, and this ended the early period of the discussion of Young's theory. After this general survey I will now pass to specific British review journals.

VI.2 British reviews of Young's theory (1801–1805)

To simplify the analysis of British reviews of Young, I have divided their authors into two groups according to their approach to reviewing physical, and particularly optical, works. The first group embraces the "non-experts." These reviewers did not consider themselves proficient in the subject, and consequently their comments were short and contained only a summary of the papers' contents and a few quotations. The "non-experts" avoided debating the authors' views, and when any evaluation was presented, it was very general. The reviewers from the second group (the "experts") considered themselves competent to make specific judgements on theoretical and experimental problems. Frequently they preferred to discuss their own views rather than the author's. The reviewers of the *Annual Review, Annals of Philosophy, British Critic, Monthly Magazine*, and *Philosophical Magazine* belong to the first group, while those of the *Critical Review, Monthly Review, Edinburgh Magazine*, and *Edinburgh Review* represent the second type. I will now examine how different reviewers treated Young's papers, as compared to the works of others.

[21] [Brougham], "On Young's Bakerian lecture," 103.
[22] Ibid., 100.

a) "Non-experts"

In the first group the *Philosophical Magazine* did not present any evaluations of Young.[23] The *British Critic* said of the "Theory of Light" that "nothwithstanding Dr. Young's *ingenious and truly laudable* endeavours, the theory of light is yet very far from being sufficiently understood."[24] The *Monthly Magazine* called the same paper "a *very elaborate* learned discourse," and judged Young's propositions to be *"well-supported."*[25] It asserted that the "Production of Colours" "will be read with *interest*," and the "Experiments and Calculations" was a "*valuable* paper."[26] The *Annals of Philosophy* believed that "The *ingenious* author of the preceding paper ["Theory of light"] having established the following law [of interference] . . . has applied it in a *very satisfactory* manner to the explanation of the colored fringes of fibres, of mixed plates, and of atmospheric halos."[27] The *Annual Review* emphasized the difficulty of Young's task and the "*importance* of the propositions," believing that "the subject merits, and will receive ample discussion."[28] It also stated that Young's law of interference "is applied in a *very ingenious* manner to the colors of fibres and the colors of mixed plates."[29] Young's experiments in the "Experiments and calculations" were called "*easily made* and whose results are *clearly detailed*."[30] None of the "non-experts" directly complained about any obscurity of the reviewed papers, though sometimes they confessed to an inability to abridge intelligibly their contents.[31]

For the sake of comparison let us look at the reviews of Wollaston's paper on refraction in Iceland spar and of Jordan's books on colors by inflection and of

---

[23] "Proceedings of Learned Societies. Royal Society of London," *The Philosophical Magazine* 2 (November 1801): 76–7; ibid., 13 (1802): 289.

[24] "*Philosophical Transactions* for 1802. Part I. The Bakerian lecture. On the theory of light and colours. By Thomas Young, etc." *The British Critic* 22 (August 1803): 137. Italics added.

[25] "Proceedings of Learned Societies. Royal Society of London," *The Monthly Magazine* 13 (July 1, 1802): 567. Italics added.

[26] Ibid., 15 (April 1, 1803): 256. Italics added.

[27] "Natural Philosophy," *The Annals of Philosophy* 3 (1804): 9–10. This journal is the only one which paid attention to the optical part of Young's "Outlines of experiments." The reviewer seemingly adopted the concept of the ether. He called Young a "*very ingenious* writer" and praised his "experimental skill" and "mathematical precision"; see "Miscellaneous Articles in Natural Philosophy," *Ann. Phil.* 1 (1801): 42–43. Italics added.

[28] "*Philosophical transactions of the Royal Society of London* for the Year 1802. The Bakerian lecture. On the theory of light and colours. By Thomas Young," *The Annual Review* 1 (1802): 875. Italics added.

[29] Ibid., 881. Italics added.

[30] "*Philosophical Transactions of the Royal Society of London* for the Year 1804. The Bakerian lecture. Experiments and calculations relative to physical optics. By T. Young," *The Annual Review* 3 (1804): 871. Italics added.

[31] "*Philosophical Transactions* for 1802. Part II. An account of some cases of production of colours . . . By T. Young," *The British Critic* 23 (January 1804): 109.

thin films. Though all the reviews of Wollaston's paper were very short, several of them contained a statement about the agreement between Huygens' theory and Wollaston's observations, and none of them criticized the wave theory.[32] The evaluations of Jordan's books were also favorable. He was not censured for correcting Newton's explanations of diffraction and of thin films. The *Philosophical Magazine* commented on Jordan's explanation of inflection that, "whithout having ourselves repeated the experiments, and without knowing them to have been repeated, with similar results, by others, we would not presume to decide concerning the truth of the doctrine."[33] Thus, the "non-experts" take a "moderate" approach to scientific reviewing. They show no theoretical biases, nor do they consider it possible to make a definite judgement on a theory or observation until they have fully mastered the necessary experiments or calculations. They even admit the possibility of being mistaken. For instance, when (correctly) stating that Jordan's theory of thin films is quite obscure, the reviewer of the *Philosophical Magazine* remarks: "Perhaps, however, not the author, but the dullness or impatience of the reviewer, may here be chiefly in blame."[34] On the contrary, the "experts" consider themselves to be infallible, as we will now see.

b) The Critical Review

The reviewer does not object to the use of hypotheses – he himself uses the hypothesis of the ether pervading all bodies – but he objects to light being an undulation in the ether. He sees a decisive argument against the wave hypothesis in the rectilinear propagation of light.[35] When commenting on theories supported by Wollaston, Jordan, or Young, the author is more concerned with his own theory. Light, in his opinion, is a "chemical substance," while colors are produced by an "attenuation" of light; for instance, a condensed white light becomes diluted due to an attraction to a body and acquires color.[36] The reviewer rejects Young's ac-

[32] "*Philosophical Transactions of the Royal Society of London.* On the oblique refraction in Iceland crystal. By W. H. Wollaston," *The Annual Review* 3 (1804): 881; "Proceedings of Learned Societies. Royal Society of London," *The Philosophical Magazine* 2 (1801): 289; "*Philosophical Transactions* for 1802. Part II. On the oblique refraction of Iceland crystal. By William H. Wollaston," *The British Critic* 23 (1804): 109.

[33] "New Publications. *New Observations Concerning the Inflexions of Light . . .* (London, 1799)," *The Philosophical Magazine* 7 (1800): 366.

[34] "New Publications. *New Observations Concerning the Colours of Thin Transparent Bodies* (London, 1800)," *The Philosophical Magazine* 8 (1800–1801): 181.

[35] "*Philosphical Transactions* for the Year 1802. The Bakerian lecture. On the theory of light and colours. By Thomas Young, etc.," *The Critical Review* 35 (August 1802): 362–63

[36] "*Philosophical Transactions of the Royal Society* for 1804. The Bakerian lecture. Experiments and calculations relative to physical optics. By Thomas Young . . .," *The Critical Review* 3 (1804): 170; and *"Observations on the Inflections of Light . . ." The Critical Review* 34 (1802): 438–42.

count of the production of colored fringes by interference as inferior to his own explanation of the phenomena. In fact, Young's "Production of colours" (as well as Wollaston's paper on refraction in Iceland spar) is unintelligible to him.[37] He favors Newton's theory of light no more than Young's; in particular, he objects to the representation of light rays by geometric lines and the production of colors by the separation of luminous rays.[38] Thus, despite the author's claims to the contrary, it is clear that he was completely incompetent in physical optics and would have not understood Young even if his writing was perfectly lucid.

c) The Monthly Review

The reviews of Young in the *Monthly Review* were written by Robert Woodhouse (1773–1827), a prominent English mathematician who first introduced into England the Leibnizian notation and methods of calculus.[39] He was a fellow of Caius College, Cambridge since 1798 and became Lucasian Professor of Mathematics in Cambridge in 1820, and Plumian Professor of Astronomy and Experimental Philosophy in 1822. He was very perceptive in commenting upon mathematically formulated problems in physics. He was, for example, the only reviewer of Wollaston's paper to explain the idea of Huygens' theory of double refraction.[40] On the other hand, Woodhouse was less successful when evaluating qualitative physical theories and experiments. For instance, he became so disappointed with Rumford's vibratory theory of heat that he recommended to the author to do experiments and leave the formulation of theories to others.[41] At the same time he praised John Leslie's theory of heat, in which the matter of heat was chemically combined with particles of air and was transmitted by the air waves.[42]

With this background in mind we would hardly expect from Woodhouse a proper understanding and appreciation of Young's theory of light and colors,

---

37 *The Critical Review* 34:440; ibid, 35:363.

38 "*Philosophical Transactions of the Royal Society* for the Year 1802. Part II," *The Critical Review* 38 (May 1803): 26.

39 B. C. Nangle, *The Monthly Review. Second series, 1790–1815. Indexes of contributors and articles* (Oxford, 1955), 74, 182.

40 [Woodhouse], "*Philosophical Transactions of the Royal Society.* Part II for 1802. On the oblique refraction of Iceland crystal. By William Hyde Wollaston," *The Monthly Review* 40 (1803): 415–16.

41 [Woodhouse], "*Philosophical Transactions of the Royal Society.* Part I for 1804. An enquiry concerning the nature of heat. By Benjamin Count of Rumford," *The Monthly Review* 46 (1805): 232, 234.

42 [Woodhouse], "Leslie on nature and propagation of heat," *The Monthly Review* 45 (1804): 85–90.

although he was quite tolerant and respectful of Young. He began his review of the "Theory of light" by expressing a doubt as to the possibility of discovering the nature of light. Young, he felt, did not answer Newton's objections to the wave theory, so that

> it is not surprising if we do not find ourselves converted to the truth of his theory; nor if against particular parts of it, we should in a minute examination enter our protest. Instances also occur of gratuitous assumptions, of hypotheses too extensive, and of conclusions not strictly logical; but . . . curious and philosophic will *do well* to peruse it, and jugde for themselves.[43]

Thus, despite the apparent inadequacy of Young's arguments, Woodhouse did not dismiss his theory altogether, as he did with Rumford. Although he did not say it directly, Woodhouse's opinion of Young's criticism of Newton was probably similar to his view of Jordan's anti-Newtonian stand:

> To find defects in the accurate philosophy of Newton, or to make opposition to the authority of so great a name, appears to be a difficult and dangerous untertaking: yet the author of the present treatise has no cause for regret at the exertion of his talents in such an attempt. His observations will be found to be made with judgement, and expressed with great modesty; if, therefore, he may not escape the hostility of the advocates of the Newtonian doctrine, he at least ought to be secure from the severity of repulse which a wanton, a petulant, and an imbecile attack on a long established system might deservedly call forth.[44]

Woodhouse knew how difficult it is to reject an established view, since in his fight against Newton's method of fluxions he himself became a "rebel." Apparently, he respected Young's stand.

As Woodhouse proceeded through the sequence of Young's papers his criticism lessened. In his review of the "Production of colours" he did not reject anything, and only suggested that Young adopted a clearer style of writing, for "we do not apprehend with sufficient distinctness the phenomena described, and their explanation according to the undulatory theory, so as to make them the

[43] [Woodhouse], "*Philosophical Tansactions of the R.S.* Part I for 1802. The Bakerian lecture. On the theory of light and colours. By Thomas Young . . .," *The Monthly Revies* 39 (December 1802): 405, italics added.

[44] [Woodhouse], "*New Observations on Thin Transparent Bodies*. 1800," *The Monthly Review* 35 (1801): 417.

subject of critical examination."[45] In his third review, of the "Experiments and calculations," Woodhouse presented no criticism at all. It seems that his understanding of Young's papers improved.[46] The third comment is the most significant for our purpose. Here Woodhouse presented an explanation of Young's principle of interference:

> suppose, with the hypothesis of the undulatory propagation of light equal waves of light to arrive at a certain point, and there to meet other waves, equal, and moving with the same constant velocity; then these waves must interfere with each other after this manner: if the elevations of one series coincide with those of the other, greater joint elevations are the consequence: but if the depressions of the one correspond with the elevations of the other, a counteraction takes place, and in the resulting effect neither elevation nor depression occurs.[47]

This account is even more lucid than Young's. Woodhouse was the only British reviewer who tried to grasp the meaning of Young's principle and its application. Possibly, he understood the production of colors by interference of white light, for he said that, "If the lengths of the paths described by two portions of light be equal, then color disappears, or the light is white."[48] He evaluated Young's experiments as *"simple and easily executed."*[49]

Thus, Woodhouse did not mind well-prepared anti-Newtonian attacks. As an "emissionist," he rejected Young's wave theory for leaving Newton's objections unanswered. Nonetheless, he respected Young's work and attempted to grasp his ideas.

d) The Edinburgh Magazine

David Brewster contributed to the *Edinburgh Magazine* since 1800 when he received his M.A. from the University of Edinburgh. He was interested primarily in astronomy, physics, and the methodology of science, where he defended the

---

45 [Woodhouse], "*Philosophical Transactions of the R.S.* Part II for 1802. An account of some cases of production of colours . . . By Thomas Young . . .," *The Monthly Review* 11 (April 1803): 416.

46 Woodhouse said that he understood Jordan's second treatise better than the first, not because it was more clearly written but "because we are more accustomed to his manner and his ideas" (*Mon. Rev.* 35 (1801): 279). Possibly, the same thing happened with Young's works.

47 [Woodhouse], "*Philos. Transactions of the R.S.* Part I for 1804. The Bakerian lecture. Experiments and calculations relative to physical optics. By Th. Young," *The Monthly Review* 46 (1805): 226.

48 Ibid.

49 Ibid. Italics added.

experimental method against hypotheses. He also reviewed books and published miscellaneous notes on science. In July 1802, Brewster noted that,

> Dr. Young . . . has *revived* in part the Theory of Light, first broached by Huygens . . . From a series of propositions, in some of which he attempts an explanation of many of the phenomena of light; he draws this general conclusions, "that radiant light consists in undulations of the luminiferous Aether," and endeavours to obviate the objections which may be raised against his Theory.[50]

For some reason Brewster commented on Young's theory not here but a page further, while discussing Thomas Pownall's paper on colors.[51] Pownall suggested that only one primary color exists, with all others being gradations of this color towards pure light or darkness. Brewster remarked in a footnote that

> It was the opinion of Sir Isaac Newton that there were seven distinct primary colors, differing from each other in refrangibility, and that by the union of them a pure white light was formed. The celebrated Tobias Mayer . . . supposes, on the other hand, that there are only *three* primary colors, viz. *red, yellow*, and *blue*, because from the mixture of these, every other colour may be produced. These different opinions arise from different definitions of a primary colour given by these philosophers. The opinion of Mayer is more simple and popular, and that of Newton more philosophical and just. That of Governor Pownall is as perplexing as it is new. Why the degradations from light towards its actual absence should be blue, indigo, and violet, is a circumstance rather unintelligible. The opinion, we presume, is as *hypothetic* as the theorem on which it is founded.[52]

Thus, Brewster discusses the number of primary colors and juxtaposes Newton's and Mayer's opinions, which have some support, to a purely hypothetical view of Pownall.[53] In the next sentence Brewster mentions Young:

---

[50] D[avid] B[rewster], "Literary and scientific notices," *The Edinburg Magazine* 20 (July 1802): 63, italics added.

[51] Pownall, "Inquiries into colored light . . .," *The Philosophical Magazine* 12 (1802): 42–49, 107„112.

[52] *The Edinburgh Magazine* 20: 64, my emphasis in "hypothetic."

[53] In fact, it was the unfortunate way of reasoning rather than the lack of observations, that led Ponwall to erroneous conclusion.

> And we have some foundation for believing that neither the hypothesis of Pownall, nor that which is newly *revived* by Dr. Young, will ever overthrow the beautiful and simple hypothesis of Newton.[54]

The context of the passage suggests that Brewster is comparing Newton's theory of colors with Young's theory of color vision. In fact, Brewster never fully understood either of these theories, but this is of no significance for my purpose, for here I am interested solely in establishing whether Brewster was referring to Young's theory of color vision. In my view, the word "revived" excludes this possibility. The idea of the analysis of light by three receptors sensitive to different colors was Young's discovery, while the belief that every color may be obtained by a mixture of three primary colors was widespread, especially among artists, and was supported by some scientists (Tobias Mayer, for instance). Therefore, as his next sentence shows, Brewster was referring to Young's revival of the wave theory:

> There is one insuperable objection to the undulatory system to which its *reviver* Dr. Young paid no attention. This system cannot explain the hyperbolic fringes which Newton mentions in 10th Observation of the third book of his Optics, or, rather the very existence of these fringes is an irrefragable argument in favour of the hypothesis which Newton framed.[55]

Now a question arises: how does Brewster pass from a theory of colors to the wave theory of light? Brewster apparently alludes to Young's principle of interference (an undulatory concept to him) as a new explanation of the formation of colors. Since in Young's theory a mixture of two white beams can form a color, Brewster considers this to contradict Newton's concept of white light as a mixture of simple colors. Not only is the fact of this misunderstanding interesting, but so is Brewster's judgement of the persuasiveness of the principle of interference: he compares this principle with Pownall's idea, which is for him a pure hypothesis.

Brewster's interest in the explanation of diffraction is evident in his own studies. In 1798 he experimented on how the nature of the material of diffracting bodies influences the positions of the fringes, but he did not publish his results

---

[54] *The Edinburgh Magazine* 20:64, italics added.

[55] Ibid., italics added.

until 1817.[56] His reasons for believing that hyperbolic fringes are incompatible with the wave theory are obscure. It is rather surprising that he did not comment on Young's study of diffraction in the "Production of colours" and "Experiments and calculations." In the period before 1807, he mentioned Young only once, in his comments on Wollaston's paper on refraction in Iceland crystal.[57]

e) The Edinburgh Review

In the early years of the *Edinburgh Review* Brougham was their only reviewer of science, and optical treatises were one of his major interests. He was a competent scientific observer, and definitely the best among British reviewers of optical works. He was especially acute in discussing experimental works. For instance, Brougham recognized the importance of a short note in Wollaston's paper that announced the discovery of ultraviolet light, whereas other reviewers ignored it.[58] While praising Leslie's experiments on heat, he correctly objected to his theory of propagation of heat.[59] Brewster and Giovanni Battista Venturi (1746–1822), and later Arago and Baden Powell (1796–1860) favorably referred to Brougham's optical works.[60]

Brougham used his reviews to promote his own views on science, in general, and optical studies, in particular. He advocated a purified and improved Newtonianism, as he understood it, and opposed to hypotheses, such as Newton's theory of fits, the ether, and the wave theory of light. Young was not his only target. Brougham accused James Wood of devoting too much atention to Newton's theory of fits.[61] He admonished Wollaston for embracing "the *wild* optical theory of vibrations."[62] He called Rumford's vibration theory of heat a "*fanciful* theory," filled with "many *absurdities*,"[63] and compared this theory with that "which Swift

[56] "Royal Society of Edinburgh," *Quarterly Journal of Science* 2 (1817): 207. Brewster noted that the independence of diffraction on the substance of a diffracting body contradicts Newton's explanation of diffraction. There is no evidence that Brewster held to the same anti-Newtonian stand in 1798, as Morse claims; "Natural philosophy," 83.

[57] D[avid] B[rewster], "Progress of manufactures, science, and fine arts," *The Edinburgh Magazine* 21 April 1803): 20.

[58] [Brougham], "Wollaston on prismatic reflection," *The Edinburgh Review* 2 (1803): 98.

[59] [Brougham], "Leslie's inquiry into the nature of hat," *The Edinburgh Review* 7 (1805): 63–91.

[60] See n. 55, above, and also [Brougham], "Venturi, Sopra i Colori," *The Edinburgh Review* 7 (1805): 27; Fresnel, *Oeuvres* 1: 6; and Powell, "Further remarks on experiments relative to the interference of light," *The Philosphical Magazine* 1 (1832): 433.

[61] [Brougham], "Wood's *Optics*," *The Edinburgh Revies* 1 (1802): 62.

[62] [Brougham], "Wollaston on Iceland Crystal," *The Edinburgh Review* 2 (1803): 99. Italics added.

[63] [Brougham], "Count Rumford on the nature of heat," *The Edinburgh Review* 4 (1804): 400, 409. Italics added.

has ridiculed in his picture of the Laputan projector, who wasted his life in extracting sun-beams from cucumbers, in order to preserve the rays for use during winter."[64]

Thus, we see in these instances that the style of Brougham's reviews of Young was not reserved exclusively for Young. There was, however, one feature of his criticism of Young which was indeed unique. When commenting on Wood, Wollaston, Rumford, and others, Brougham never failed to emphasize their positive contribution. For instance, he praised Rumford as an "*ingenious and persevering* observer," and Wollaston as an "*acute and ingenious* experimentalist."[65] On the contrary, Young was for him neither a philosopher nor an experimentalist, and his papers were "destitute of every species of merit."[66] It had been generally accepted in the past that Brougham's criticism of Young was to a large degree inspired by personal animosity. Recently, however, some scholars have maintained that Brougham was primarily concerned with defense of the true methodology of science.[67] Without denying the methodological aspect of Brougham's reviews, I intend to show that his discussion of Young's experiments clearly demonstrates his personal bias against Young.

Brougham's general attitude toward Young's experiments may be characterized by his comment on Young's hair experiment: "The new case of colors . . . has been observed a thousand times; and he [Young] has only the merit of giving an absurd and contradictory explanation."[68] Let us follow the order in which Brougham himself discusses Young's observations.

In the case of the internal fringes Brougham concentrates on the "absurdity" of Young's explanation, but he does not indicate any predecessor of Young. According to Brougham, the observed fringes are due to refraction and not to interference of inflected and reflected light. He asserts that, due to transparency of water drops, a halo around the sun or the moon results from refraction and not diffraction, as Young claimed. Subsequently, Verdet showed, however, that the

64 Ibid, 411.

65 [Brougham], *The Edinburgh Review* 2 (1803): 99; and 4 (1804): 413. Italics added.

66 [Brougham], "Dr. Young's Bakerian lecture," 450.

67 Cantor, "Henry Brougham and the Scottish methodological tradition," *Studies in History and Philosophy of Science*: 69–89; Steffens, *Newtonian Optics*, 133; and Morse, "Natural philosophy," 107–11.

68 [Brougham], "Young on colours," 457.

primary cause of the phenomenon is diffraction.[69] To support his point on the role of refraction, Brougham refers to Young's failure to observe fringes with a "black" hair (he misread "horse" hair) as a proof that opaque bodies do not produce fringes. His reasoning apparently is as follows: when there is no refraction, there are no fringes; therefore, the fringes are produced by refraction. Brougham ignores the fact that in those cases when Young saw the fringes, the hairs or threads were also opaque.

The colors of "mixed" plates are for Brougham just a repetition of Newton's observations of the colors of thin films; and he explains the difference in the fringes' width in the two cases by the fact that Young's experiment was simply an "inelegant" form of Newton's.[70] In other words, he hints at an inaccuracy in Young's observation. Brougham also accuses Young of misinterpreting the transformation of the black spot at the center of Newton rings into a white one, claiming that it has nothing to do with the intermediate liquid, but is caused by an increase of the distance between the lenses.[71] Of course, by failing to mention that he kept the pressure on lenses constant at different stages of this experiment, Young himself created a possibility for misunderstanding. Brougham, however, is obviously too quick in judging Young to be a careless experimentalist. He also commits a similar mistake in commenting on the "screening experiment." In Brougham's view, the screen *PO* (Fig. 22) was moved too far into the shadow and intercepted *both* interfering rays instead of one.[72]

When Brougham was unable to deny the phenomena or invent an error in Young's account he simply dismissed Young's explanation and proposed an alternative of his own. This happened with the colors of flames and with microscopic observations.[73] Unlike his other reviews, when commenting on Young Brougham did not show himself an able experimentalist. The reason for this is clear: he was not interested in an objective study of Young's experimental work; he was just looking for errors, even if he had to invent them. Reviewers were not expected to repeat the experiments in the works under review, and consequently they were usually quite reserved in their judgement about them. Brougham him-

---

[69] Ibid. Young's explanation is the only possibility in the two-ray model, and it is quite satisfactory. Some historians support Brougham's claim that Young's account of halos was wrong because this phenomenon can be explained by refraction without any interference; Morse, "Natural philosophy," 113, 115; and Cantor, "Thomas Young's lectures at the Royal Institution," *Notes and Records of the Royal Society of London* 25 (1970): 102. They are confusing two kinds of halo, one of which has a constant diameter (22° or 47°) and is produced by refraction, and another of a variable diameter and due to diffraction. Young discussed both types; see, for instance, his *Lectures* 1:443, 466. On Verdet's theory see n. 70, Ch. V

[70] Ibid.

[71] Ibid., 458.

[72] [Brougham], "Dr. Young's Bakerian lecture," 99.

[73] [Brougham], "Young on colours," 458–59.

self was more careful when commenting on experiments of Wollaston, Rumford, Leslie, and Venturi, but Young was an exception. Whatever the reason for Brougham's antipathy to Young, the fact itself is established.[73a]

Brougham's discussion of the principle of interference shows his misunderstanding of the subject. Unlike Woodhouse, Brougham is unwilling to admit the wave theory, even if temporarily, in order to pursue the consequences that Young claimed followed from it. Thus he attempts to find a mechanical justification for the principle of interference in the emission theory by means of an interaction of two beams of light particles. Since his emission theory does not incorporate any periodicity, he claims that the result of interference of two rays could yield only one fringe and no more.[74] He then argues that it is "mathematically impossible" for rays of light to be inflected and reflected as much as Young required.[75] This argument is meaningless without assuming a particular law of attraction between light and matter and a particular incidence of reflected light. For Brougham, the most incomprehensible part of Young's theory was the production of colors. He exclaims:

> ... upon what known principle of optics can it be conceived, that the very cause of whiteness, a mixture of rays, should create colour, and that two beams uniting, in what way or by what new laws soever, into one beam, should, by that union, become separated into several coloured fringes, with dark intervals and a white centre?[76]

When comparing this passage with the one quoted above (n. 103, ch. V) in the discussion of the experiment with two candle flames, we see that Brougham is concerned there not with the problem of coherence but with the formation of colors. He believes that the principle of interference is incompatible with Newton's theory of colors, where a colored ray is the primary concept, and a mixture of colored light produces whiteness, while Young seemingly claims just the opposite, namely that white light is the primary concept and a mixture of two portions of white light produces colors. Apparently, in the same passage Brougham also uses another argument against the interference of light, namely that the color of light does not depend on its intensity. Since a mixture of two portions of

[73a] Many years later, Brougham mentioned Young in his autobiography (as his "fellow student under Black") with the following footnote: "Thomas Young, celebrated for deciphering the Rosetta inscription, born 1773, died 1829" (*The Life and the Times of Henry Lord Brougham, written by himself*, 2 vols, 2nd ed. (Edinburgh: W. Blackwood & Sons, 1871), 1:66)

[74] [Brougham], "Dr. Young's Bakerian lecture," 98.

[75] [Brougham], "Young on colours," 457.

[76] [Brougham], "Dr. Young's Bakerian lecture," 99.

white light only increases its intensity, it cannot, in Brougham's view, form any color. He ignores Young's requirement that the interfering portions of light must travel slightly different routes to create colors. Therefore, Brougham refuses to discuss the principle of interference according to Young's model and conditions, and after finding that this principle contradicts his own model, he rejects it as "one of the most incomprehensible suppositions that we remember in the history of human hypotheses."[77]

It is manifest from Young's "Reply" that he was afraid that Brougham's attack will affect his reputation as a physician. This explains why henceforth he avoided public debates and discussed the wave theory and principle of interference primarily in anonymous reviews. This circumstance could, of course, have delayed, to some degree, the spread of his ideas among scientists; however, there is no evidence for this. Brougham's criticism was not without its positive aspect too, because it stimulated Young to carry out further experimental investigations, reflected in his "Experiments and calculations," and also to think more about the problem of coherence.

VI.3 Later response (1807–1815)

The large number of comments on Young's optical papers in British review journals should not lead one to believe that his ideas attracted great interest, for his articles were reviewed simply as a part of the proceedings of the Royal Society of London or of the entire volumes of the *Philosophical Transactions*. Young believed that the major reason for lack of support for his theory was an unfamiliarity with it. Consequently, he did his best to publicize his works: he reprinted (with changes) his optical papers in his *Lectures*, referred to them in his anonymous reviews of other authors, and even commented on his own papers (such anonymous self-reviews were not unusual in Britain at the time).[78] Despite these efforts, very few comments on Young's optical works appeared between 1807 and 1811. Improvements in the presentation of the wave theory of light and the principle of interference given in his *Lectures* passed almost unnoticed. Two re-

[77] Ibid., 97.

[78] There is internal evidence that Young is the author of various comments on his own works, among which are the reviews of his optical papers in the *Imperial Review* (see 2 (1804): 247–48, 257; and 3 (1804): 428–30), two short notes on the forthcoming *Lectures* (*Nicholson's Magazine* 18 (1807): 79–80, and *The Monthly Magazine* 20 (1805): 57) and a large review of the *Lectures* (*The British Critic* 30 (1807): 1–15). It is quite evident that the review of the *Reply* (*The British Critic* 25 (1805): 95–97) belongs to Young, and it is possible that a short note on the *Reply* (*The European Magazine* 47 (1805): 130–31) was also written by him.

viewers evaluated the *Lectures* positively, but neither was interested in optics. One simply noted that Young placed optics in the section on hydrodynamics in order to show his predilection for the wave theory of light,[79] and the other did not at all mention optics.[80] In 1808, Richard Winter published a note on the "specific gravity of light" in which he defended Young's wave theory.[81] However, in his explanation of diffraction he used a speculation of his own, and not the principle of interference. Young was not happy with such a support and rejected it.[82] In 1809, William Nicholson (1753–1815), a chemist and editor of the *Journal of Natural Philosophy* emphasized that "*many* philosophers" had turned to the wave theory of light and that "among modern philosophers who have supported this doctrine, Dr. Young has shown *much ability* in his experimental and theoretical researches."[83]

In 1811, Ludwig Wilhelm Gilbert, Professor of Physics at the University of Leipzig and the editor of *Annalen der Physik*, asked August Friedrich Lüdicke (1748–1822), a professor of mathematics in Meissen, to translate all Young's optical papers for his journal.[84] Gilbert, Lüdicke, and Karl Brandan Mollweide (1774–1825), Extraordinary Professor of Astronomy at the University of Leipzig, briefly commented on them. While complaining of Young's obscurity, they understood, nonetheless, some parts of his works. Mollweide correctly reconstructed Young's derivation of the path difference for the external fringes produced by a hair in Newton's Observation 3 and derived exactly the same wavelength as Young did. He also grasped the mathematical content of the principle of interference, but did not adopt it as a physical concept, for he considered it to be a "quite arbitrary hypothesis."[85] Young's idea that colors could be formed by a mixture of two pencils of white light appealed to Lüdicke, who claimed to have demonstrated that colored fringes and the whole prismatic spectrum could be made from a mixture of two diffracted light rays.[86] In 1810, Lüdicke proposed a new theory of dispersion in which only those rays could produce prismatic colors which were first diffracted in a small hole. The diffraction broadens the beam of light, and rays fall on the prism at different angles and travel through different thicknesses of glass; for example, the rays passing through the thinnest

79 "Young's *Lectures on Natural Philosophy*," *The Critical Revies* 12 (1807): 7.

80 "Science and Experimental Philosophy. Young's *Lectures*," *The Annual Review* 7 (1808): 673–78.

81 Winter, "A method of finding the specific gravity of light from analogy: and the undulatory system defended by an experiment in inflected light," *Nicholson's Magazine* 19 (1808): 143–46.

82 *Retrospect* 4 (1809): 36–38.

83 [Nicholson], "Light," *The British Encyclopaedia* [1809]; American edition: vol. 7 (Philadelphia, 1818), there ist no pagination.

84 *Annalen der Physik* 39 (1811): 156–205; 206–20; 255–61; 262–91.

85 Ibid., 285–87.

86 Ibid., 284–85.

part are yellow and red, while those which pass through the thickest part are blue and violet.[87] Lüdicke said that Young's principle of interference was valid, but only for diffracted light, that is, for light passing by the edge of a hole and not for light travelling undisturbed through the middle of the hole. Thus, Lüdicke treated the principle of interference as a non-Newtonian theory of colors. An opposition to Newton's theory of colors was widespread in Germany, and German scientists apparently misunderstood Young as their ally in this question. The controversy on the nature of colors became more agitated after the publication of Goethe's *Farbenlehre* in 1810.[88] It is likely that this event accelerated the publication of Young's papers promised by Gilbert as early as 1806.[89]

Another stimulus for interest in Young's works was Malus's discovery of polarization of light, which prompted physicists to turn to optics and make it one of the most popular branches of physics in the second decade of the nineteenth century. It must be emphasized that this new interest in Young's works was connected not with his wave theory, which, in Young's own opinion, was insufficient to explain the phenomena of polarization,[90] but with the principle of interference and Young's experiments. In 1812, Davy, who thought that the new discovery favored the emission theory, expressed a hope that periodical colors also will be accounted for on the basis of Newton's "poles" and Malus' "axes" in light particles:

> May not the experiments of Dr. Young, *Phil. Trans.* 1804, page 2, which he considers as proving that homogeneous light at certain equal distances, in the direction of its motion, is possessed of opposite qualities capable of neutralizing each other, and of extinguishing the light when they happen to be united; be explained on the idea of attractive poles in opposite sides of the particles of light. That able philosopher considered them as favourable to the theory of undulation; but if the attractions of other matter can destroy the motions of light, as in the case of its action on black bodies, may not the same result be produced by the attractions of its particles for each other?[91]

87 Lüdicke, "Versuche die Mischung prismatischer Farben," *Annalen der Physik* 4 (1810): 232–35.
88 See for istance, Mollweide, *Prüfung der Farbenlehre des Hrn. von Goethe und Vertheidigung des Newton'schen Systems gegen dieselbe* (Halle, 1810). I have not seen this work.
89 Another factor which prompted Ludicke to translate Young's papers in 1811 might have been his own work of 1810.
90 [Young], "Review of the *Memoires d'Arcueil* [1810]," 248–49.
91 Davy, "Elements of chemical philosophy [1812]," *Coll. Works* 4: 161.

Although Young's articles were available to French academicians since 1803, I have found no comments on the principle of interference prior to 1811.[92] In 1812, Jacques-Étienne Bérard (1789–1869), Berthollet's assistant, who was studying polarization of ultraviolet rays, recalled Young's earlier experiment which showed that these rays produce Newton's rings as visible rays do.[93] At about the same time Arago referred to Young in his paper on thin films (both Berard's and Arago's papers were published only in 1817).[94] Arago studied the reflection and refraction of polarized light by thin crystals and found that both surfaces of the plate contribute to the production of colors. This reminded him that Young had also employed two surfaces to explain Newton's rings, which explanation Arago believed to be analogous to Hooke's. There is no evidence, however, that Arago adopted the principle of interference until late October 1815, when he became familiar with Fresnel's memoir on diffraction (see more on Arago in Ch. VIII). In 1815, Arago recommended to Fresnel Young's papers in *Philosophical Transactions* and apparently his *Lectures* too.[95] Étienne Malus (1775–1812) was familiar with some Young's works in 1811 and called his principle of interference "your *ingenious* hypothesis of the combined movements of light."[96]

Having been impressed by Young's explanation of "chromatic polarization," Brewster wrote to Young on July 28, 1815 that he was sending him

> an account of an optical discovery, in which I have no doubt you will be much interested, as it appears to give very great support to your opinion, that the colours produced by the action of crystallized bodies upon polarized light are referable to your theory of periodical colours.[97]

[92] The Paris Academy of Sciences obtained the *Philosophical Transactions* for 1800, 1802 and 1804 in 1801, 1803 and 1805, respectively; *Proces-verbaux* 2:386, 2:656 and 3:180.

[93] Berard, "Mémoire sur les propriété des différentes espèces de rayons qu'on peut separer au moyen du prisme de la lumière solaire [1812]," *Memoires d'Arcueil* 3 (1817): 41.

[94] Arago, "Sur les couleurs des lames minces," ibid, 228.

[95] Fresnel, *Oeuvres* 1:6–7. See VII.1.

[96] Malus to Young, June 1, 1811: Royal Society, Young correspondence, MS 242.22. That was a reply to Young's letter, in which Young, as a Foreign Secretary of the Royal Society, informed Malus that Royal Society awarded him the Rumford Medal for his work on polarization. Young also enclosed in the letter an eriometer, "as an illustration of this *general law* which I have established" (see Young to Malus, March 9, 1811: Royal Society, Ms 581.45). Since Young did not reveal in this letter the content of his "general law," Malus had to know it from other sources. Italics added.

[97] Young, *Misc. Works* 1:359.

Young answered:

> I am glad that a *circumstance has occured*, which has led you to think with some attention of my theory of the interference of light, because I am quite certain that it must become particularly interesting to you, who have observed so many phenomena which appear to be most intimately related to it, and which admit of a striking illustration by its means. I cannot, however, see that it has much connexion with the facts which you state respecting internal reflection.[98]

Thus, Brewster erred again about where and how to use the principle of interference. For this reason, Young showed him how to do it for the colors made by two identical and almost parallel plates, recently discovered by Brewster. The explanation was not very clear, but it emphasized that Brewster's phenomenon differed from the one discovered earlier by Nicholson, who used two parallel plates of almost the same thickness. By referring to the "circumstance" Young hinted that Brewster's interest in the principle of interference emerged only when he began to deal with periodical colors. It seems that by 1815 Young's wave theory had become sufficiently well known in Britain to provoke a detailed refutation of it by William Crane.[99]

In Frankel's view, the French attitude towards Young was expressed by Biot in his letter to Brewster of February 20, 1816:

> I very much esteem the merit and talent of this distinguished scientist, but you will permit me to tell you that his witness has no more weight here than the authority of Aristotle against the observations of Galileo

[98] "Dr. Young to Sir Dr. Brewster, September 13, 1815," ibid., 360–62.

[99] Crane, "Observations upon the different hypotheses that have been proposed respecting the nature of light," *The Philosophical Magazine* 46 (1815): 195–203.

on motion... Even if his theory, *which I do not know at all* were true... [emphasis is mine].[100]

The first sentence seems to imply that Young had been treated in France as a speculative philosopher rather than an experimenter, while the second one hints that Biot knew nothing about Young's wave theory. As to the former, we should not rush with giving too much weight to Biot's appraisal of Young until we know the reason for it, since as shown further, sometimes Biot's judgements were groundless.

The latter hint, however, is not Biot's but Frankel's. It is clear from the context of the letter that the "theory" Biot is referring to is not the wave theory of light in general, but a very specific theory of the colors of double plates.[101] In his preceding letter to Biot of February 2, 1816, Brewster claimed the priority in discovering the colors produced by two identical plates, which Biot attributed to Nicholson.[102] Brewster argued that Nicholson's experiment was different from his and quoted Young's explanation. The appeal to Young's authority provoked the sarcastic remark just cited, and Biot displayed no interest either in Brewster's arguments nor in reading Young's theory of both phenomena and insisted that his own experiments showed that the equality of the plates' thickness played no role.

Biot displayed similar arrogance towards Young's theory of chromatic polarization:

> I must confess that I was not a little astonished to find that the deduction seemed to make no impression whatever on Biot, when I sent it him: for in a very civilly intended answer, he merely referred me to his own book, which I have not read, and which I believe I shall not read, because I find that the begins with suppositions respecting the motions of the particles perfectly incompatible with the general laws of mechanics.

---

[100] Frankel, "Corpuscular optics and the wave theory of light," *Social Studies of Science* 6 (1976), 141–84 (155).

[101] j'estime beaucoup le mérite et le talent de ce savant distingué mais vous me permettrez de vous dire que son témoignage n'a pas plus de poids ici que n'en avait l'authorité d'Aristote contre les observations de Galilée sur la pesanteur. M. Young ne peut rien contre des faits tels que ceux que je viens de vous citer. Il parait croire que l'egalité de vos plaques est nécessaire pour la production du phénomène et il explique cette nécessité d'apres la thèorie, mais il y renoncerait sans doute s'il savait qu'on produit tant aussi bien les anneaux avec les plaques d'inegales épaisseurs. Même si sa thèorie, que je ne connais point, est exacte il doit reconnaître a priori que cette égalité n'est nullement nécessaire. Dans ce cas je désirerais qu'il nous resolut ce problème: etant donné l'épaisseur de chaque plaque, leur inclinaison mutuelle, la direction des rayons incidens que je suppose simples et enfin la position de l'oeil, quel doit etre l'intervalle des bandes lumineuses. (Biot à Brewster, 20 fevrier, 1816: MS 4895, no. 70, fol. 2).

[102] Brewster to Biot, February 2, 1816: Bibliothèque de l'Institut, Ms. 4895, no. 69.

> I can only suppose that I have not made myself sufficiently intelligible to him.[103]

Unlike Brewster, by 1816, Biot was an expert on periodical colors in general, and the highest authority on chromatic polarization. Apparently, his ego did not allow him to admit that Young's four-page obscure theory of this phenomenon (against almost four hundred pages of his own theory) was worth of wrecking his brains over it. One may notice that Young responded in kind.

It is dufficult to imagine that Biot was totally unfamiliar with Young's theory, taking into account that not only Young's optical papers were available at the Academy of Sciences but Biot obtained some of them personally from Young when they met in Paris in the summer of 1802.[104] Probably Young made a good impression on Biot, since the latter assisted in electing Young a corresponding member of the *Societé Philomatique*, of which he was the secretary.

Although no public comments on Young's theory appeared in France before 1816, it seems that French physicists did respond, although in a specific way: by extending the emission theory to new phenomena, such as atmospheric refraction (Laplace, 1805) refraction of gases (Biot and Arago, 1806), double refraction (Malus and Laplace, 1809), and chromatic polarization (Biot, 1812). This development of the emission theory has already been described by historians, but I would like to raise a new point. Although it had no direct relation to their emission theories of particular phenomena, their authors tended to bring forth the question of the nature of light. Laplace, for instance, tested both the emission and the wave hypothesis in discussing the possibility of a resistance to planetary motion.[105] When Biot and Arago found that the refracting power of a compound substance depends on the refracting powers of its components they concluded that this result favored the emission theory and contradicted the wave theory.[106] Malus pointed out that polarization of light was incompatible with the wave hypothesis.[107]

If we now recall that by the end of the eighteenth century physicists stopped comparing the emission theory to the wave theory, the revival of the debate on the nature of light in original works published after 1800 may be viewed as a response to Young, not to the specific issues raised by him but to the overall challenge of the wave theory.

---

[103] See n. 98, p. 362.

[104] Biot to Young, December 11, 1802: Royal Society, Young correspondence, Ms. 242.

[105] Laplace, P. *Traité de mécanique céleste* v. 3 (Paris, 1802; reprint: Bruxelles, 1967), 296–303; v. 4 (Paris, 1805), 313–325.

[106] Biot et Arago, "Sur les affinités des corps pour la lumière...," *Mém. Acad.* 1806, 301–387 (344).

[107] Malus, E. "Théorie de la double réfraction," *Mém. Sav. Étrang.* 2 (1811), 303–508 (446–47).

VI.4 Summary

Between 1801 and 1816 Young received one positive comment on his wave theory and the principle of interference (Wollaston), one intelligent account of his formulation of this principle (Woodhouse), one reconstruction of the mathematical derivation of the path difference (Mollweide), and one suggestion about the use of this principle (Brewster). No one grasped the principle of interference sufficiently well to apply it to new phenomena or to discuss Young's experiments. There is no evidence that anyone even tried (at least, in the early period) to repeat Young's experiments. It is unlikely that this "indifference" to the principle of interference was caused by a hostility to the wave theory, as has been supposed. My analysis shows no such a hostility, at least in Britain. Although British scientists considered Young's arguments insufficient to convert them to "undulationists," they did not object to any attempt to bring more clarity to the problem on the nature of light. Latchford believes that British reviewers sided with Brougham. In my view, just the opposite is true, namely, Brougham's hostility toward Young and the wave theory was unique, because the general attitude of the rest of the reviewers was neutrality, respect, and encouragement. Although they could not fully understand Young's reasoning, they suspected that there was something significant in his works; and only Brougham treated them as sheer nonsense. Some historians have suggested that Young's theory had some initial support, but was neglected after Brougham's criticism.[108] This opinion is incorrect, and interest in Young's theory in fact *increased* with time. Reviewers of Young's papers frequently complained of their obscurity, but they seldom mentioned gratuitous assumptions. All, except for Brougham, praised Young's experiments, and none of them (including Brougham) expressed disappointment at the insufficient agreement between the theory and observation. Therefore, the opinion that the internal flaws of Young's theory were responsible for its lack of acceptance is false. Apparently Young's contemporaries had different criteria for adopting a physical theory than some scholars.

There is no doubt that the terseness of Young's style severely limited the number of readers who could understand him. However, I am primarily concerned in this research with the *development* of the principle of interference and not with

[108] Silliman, "Augustin Fresnel," 112; and Steffens, *Newtonian Optics*, 135. Steffens's claim that Henry Englefield supported Young (in his paper "Experiments on the separation of light and heat by refraction. In a Letter from Sir H.C. Englefield . . . to Thomas Young . . . From the Journal of the Royal Institution, p. 202," *Nicholson's Journal* 3 (1802): 125) is based on a misindentification of the person of whom Englefield is speaking. His "Doctor's" referred to W. Herschel and not to Young.

its dissemination. To apply Young's ideas and advance them further only a few scientists were needed. As shown above, those who were proficient in mathematics and physics and possessed a strong motivation to concentrate on Young's papers were able to decipher them, at least partially (Woodhouse, Mollweide). I must also mention here Joseph Fraunhofer (1787–1826) who based his 1821 theory of the diffraction grating on Young's two-ray interference model rather than on Fresnel's theory.[109] Actually, Fraunhofer was the first who applied Young's theory. Thus, the obscurity of Young's works was not the crucial factor that prevented scientists from adopting his theory. It seems that there were two other factors responsible for this which have not yet been examined.

One of them was a misunderstanding of the principle of interference as a new theory of colors, which prompted its refutation by Newtonians (Brewster, Brougham) and support by anti-Newtonians (Lüdicke). Another factor was the mathematical character of Young's theory. The principle of interference posed a dilemma to his contemporaries: they could accept it either as a part of the wave theory or as an empirical principle. In the first case, the principle would have received a sound physical foundation, but no one was willing to accept the wave theory until it could explain *all* optical phenomena, particularly the rectilinear propagation of light and dispersion. In the second case, one could use the emission theory of light and thereby greatly increase the likelihood of the incorporation of the principle of interference into optics. This explains why, Young promoted the wave theory of light and the principle of interference independently of one another. He foresaw a delay in the establishment of the true theory of light and decided to receive a recognition for the principle of interference as a mathematical law derived from observations. However, in this case another dilemma arose: does the principle need a mechanical model or can it be treated simply as a mathematical law? The emission theory did not allow for such a model, for it could not explain, for instance, why some rays are inflected at very large angles or why interfering rays meet in more than one point. That is why Brougham, Mollweide, and others considered the principle to be an arbitrary hypothesis and rejected it. If, on the other hand, the principle of interference is adopted as a mathematical principle, then the physical cause of interfering rays (for instance,

[109] Fraunhofer, "Nouvelles modifications de la lumière, par l'influence réciproque et la diffraction des rayons lumineux, avec l'examen des loix de cette modification," *Astronomische Abhandlungen* 2 (1823): 111–12. Fresnel is mentioned here but without any specific reference, and it is possible that Fraunhofer did not read any of his papers. He referred to Young's articles in the *Annalen* for 1811; "Kurzer Bericht von den Resultaten neurer Versuche, über die Gesetze des Lichtes, und die Theorie derselben," *Annalen der Physik* 74 (1823): 358–9). Fraunhofer applied the principle of the interference in a way similar to Young's in his 1801 account of a grating. Since Fresnel's theory of a grating was not yet published, Fraunhofer probably based his account on Young, and thus understood him.

forces which inflect or reflect rays in a given direction) is of no concern. Only the proper selection of interfering rays (and some other assumptions mentioned in Ch. V) in order to obtain an agreement between the theory and observations is at issue. That was how Young himself considered his principle. However, such a mathematical approach was not popular at the time. Most scientists who responded to Young's theory were "qualitative physicists," and they were greatly concerned with mechanical models of optical phenomena. Physical optics in the early nineteenth century was essentially a qualitative science, and this was especially noticeable in the study of periodical colors.[110]

Therefore, the mathematical character of the principle of interference initially was an obstacle to its acceptance. The situation changed, however, in the 1810's when physical optics was transformed into a quantitative science by the efforts of Arago, Malus, Biot, and Brewster. Accordingly, at that time we see a growing interest in Young's works, first from Arago and Malus. Young's successful explanation of some cases of chromatic polarization in 1814 added more weight to his principle of interference and attracted the attention of Brewster, who was searching for an alternative to Biot's theory. A true interest in the principle of interference was exhibited only by the "quantitative physicists", and only if they studied periodical colors (Arago, Brewster), but not if they had their own theories of these phenomena (Brougham, Biot). As to the acceptance of the principle of interference, even at the lowest level (positive comments), by October 1815, only Wollaston and Brewster did. As the example of Fraunhofer proves, the principle of interference could be fully accepted without Fresnel's intervention; however, as shown further, Fresnel greatly accelerated this process.

Apparently, the response to Young's wave theory, whether explicit or implicit, was much richer than that to the principle of interference. This may support Cantor's assertion that many "missed" that part of Young's theory. That does not concern, however, some French "emissionists", who were preoccupied not with criticizing the wave theory but with building its emission alternative.

---

110 See n. 36, Ch. IV; nn. 24, 49, 64, 101, Ch. V; n. 87, above. Also see [Jordan], *New Observations Concerning the Colours of Thin Transparent Bodies* (London, 1800); W. Herschel, "Continuation of experiments . . ., *Phil. Trans.*, 1809: 259–302, and "Supplement . . .," ibid., 1810: 149–77; Mollweide, "Über einige prismatische Farbenerscheinungen ohne Prisma, und über die Farbenzerstreuung im menschlichen Auge," *Annalen der Physik* 17 (1804): 328–27 (esp. 335–36); Brandes, "Über die farbigen Bogen, welche man zuweilen an der innern Seite des Regenbogens bemerkt," ibid, 19 (1805): 464–75; Parrot, "Drei optische Abhandlungen: Die Theorie der Beugung des Lichtes; die Theorie der Farbenringe; und über die Geschwindigkeit des Lichts," ibid, 51 (1815): 245–321; Prieur, "Considerations sommaires sur les couleurs irisées des corps reduits en pellicules minces," *Annales de Chimie* 61 (1807): 154–79; Flaugergues, "Sur la diffraction de la lumière," *Journal de Physique* 74 (1812): 125–29; 75 (1813): 16–29; 76 (1813): 142–54.

Chapter VII

Fresnel and the principle of interference

Fresnel's principle of interference *per se* received little attention from historians, who have usually limited their task to its application to diffraction and chromatic polarization.[1] Here, I intend to discuss two questions: 1) Fresnel's innovations in the principle of interference and its applications, and 2) the role of the principle of interference in the establishment of the wave theory of light (this problem will be investigated in Ch. VIII). The main difficulty in solving the second problem is the necessity to isolate the response to the principle of interference from the response to Fresnel's entire wave theory. First, we must exclude the impact of other components of his theory such as the theory of double refraction and that of reflection and refraction. Second, when examining the reception of the theory of interference, we must eliminate the influence of the Huygens-Fresnel principle and the concept of transverse waves. Only in this way will we be able to grasp how scientists reacted to the concept of interference itself.

It is comparatively easy to deal with the period from 1815 to 1818, when Fresnel's wave theory was similar to Young's. The later period, after the introduction of the Huygens-Fresnel principle (1818) and the concept of transverse waves (1821), creates more difficulties, but, nonetheless the problem still can be resolved, at least within specific intervals of time. As shown in the following chapter, before 1830 most scientists considered the wave theory to consist solely of the theory of interference in which the principle of interference was assumed to play a dominant role. Therefore, I will argue that between 1815 and 1830, it was the difficulty in understanding the principle of interference that primarily hindered the acceptance of the wave theory.

The major stages in the development of the principle of interference by Fresnel are: 1) discovery of the principle (October 1815), 2) its generalization to many waves (late 1817), and 3) discovery of all conditions of coherence (by 1822). Since little is known of how and when these discoveries were made, I will arrange my discussion according to their public presentation: 1) presentation to the Academy of Science of the first memoir on diffraction (October 23, 1815); 2) publication of this paper (March 1816); 3) publication of the prize-winning memoir on diffraction (August 1819); and 4) publication of the article "De la lumière"

[1] See Silliman, Chs. IV, V, and Buchwald, *Wave Theory*, Chs. 5–8.

(early 1822).[2] Consequently, I will distinguish three periods in the history of Fresnel's principle of interference: from October 1815 to March 1816, from March 1816 to August 1819, and from August 1819 to the spring of 1822.

It is generally accepted that Fresnel's wave theory succeeded because, in contrast to Young's, it was based on only a few hypotheses, highly mathematized, supported by precise experiments, and clearly presented. Since I found (as shown in the next chapter) that a group of scientists supported Fresnel's principle of interference before 1819, chiefly under the influence of his 1816 memoir on diffraction, the implication is that Fresnel's theory possessed these features quite early. I intend to demonstrate that that was not the case.

VII.1 First period (1815–1816)

a) Rediscovery of the principle of interference

Fresnel, a graduate of the *École Polytechnique* and of the *École de Ponts et Chaussées*, was supervising roads construction in the South of France and dreaming about inventions and scientific discoveries. He began speculating on the nature of light in 1814, leaning to the wave hypothesis. Initially, he hoped that the phenomenon of the stellar aberrations could decide between the two hypotheses, then his interests turned to polarization. Apparently, he had not done any optical experiments until the spring of 1815 when a turn of political events gave him an unexpected vacations.[3]

---

[2] Fresnel's first paper on diffraction (cited as "Frist Memoir") is "Mémoire sur la diffraction de la lumière, où l'on examine particulierement le phénomène des franges colorées que presentent les ombres des corps eclaires par un point lumineux," *Oeuvres* 1: 9–33. It was dated October 15, 1815 and was supplemented on November 10 by another paper, which will be cited as "First Supplement," (ibid, 41–60). A revised version was published in May 1816 in *Annales*; see *Oeuvres* 1: 89–122. This article will be cited as "Second Memoir" in the notes and "1816 memoir" in the text. A part of Fresnel's prize-winning memoir on diffraction was first published in July-August 1819 in *Annales*, it will be cited as "1819 memoir." The full memoir (*Oeuvres* 1: 247–382) was published in 1826, and is referred to as "Mémoire couronné." Passages from "1819 memoir" that remained unchanged will be cited from "Memoire couronne." The paper "De la lumière" (*Oeuvres* 2: 3–145) appeared early in 1822.

[3] According to Arago, Fresnel began experimenting in "the beginning of 1815" (Arago, "Fresnel," *Oeuvres* 1: 419). It is not clear whether this means before April, when he actually obtained leasure time to do research. On the beginning of Fresnel's scientific career see E. Verdet, "Introduction," in Verdet (ed.), *Oeuvres complètes de Fresnel* 3 vols (Paris, 1867–70), 1: XXVII–XLIII; Arago, "Fresnel", 112–20; Silliman, 5–46, 144–80; Buchwald, 111–118. I did not discuss the subject of VII.1(a) in my dissertation.

When Napoléon returned from Elba, Fresnel decided to join the fight against the despot. He mounted a horse and on March 23, 1815 he arrived at the headquarters of the Royalist army of the South to volunteer his service.[4] A few days later, however, he became seriously ill and returned to his residence at Nyons. Soon Napoléon reestablished his reign, and Fresnel was dismissed from service and placed under police surveillance. After a while he returned to his home village Mathieu, in Normandy, where he began his experiments on diffraction with a few crude instruments made for him by a village blacksmith. In July, Fresnel spent some time in Paris where he "renewed acquaintance with old schoolmates and so prepared himself to scientific researches."[5] One of them was Arago, his one year senior at the *École Polytechnique*. Arago gave Fresnel a list of works on diffraction of light, which included "the work by Grimaldi, the one by Newton, the English treatise by Jordan, and the memoirs of Brougham and Young, which are part of the collection of the Philosophical Transactions."[6]

On September 23, 1815 Fresnel informed Arago of the discovery of the law of diffraction:

> I think I have found the *explanation and the law* of colored fringes which one notices in the shadow of bodies illuminated by a luminous point. The results of my calculations are confirmed by the observation. However, I cannot yet bring into these observations the degree of exactitude necessary to be absolutely certain in the correctness of my equation. For this I need the instruments, which I can procure only in Paris. Before making this expense, I would like to know whether this is not useless, and whether the *law of diffraction* has not already been established by sufficiently exact experiments. Thus, I ask you, Sir, if this phenomenon has been submitted to a calculation, to let me know the equation which it represents and the theory on which it is based.[7]

---

4 M. A. Marc, *Notices sur A. J. Fresnel* (Caen, 1845), 11–12.

5 Arago, "Fresnel," 117–18. According to Arago, Fresnel arrived in Paris on his way to Mathieu ("Fresnel obtained the permission to pass through Paris," ibid). This sounds confusing because Napoleon abdicated on June 22.

6 Fresnel, *Oeuvres* 1:6. Since Fresnel received Arago's note right before his departure ("Mais je ne reçus votre billet qu'au moment même où je quittais Paris": ibid, 6–7) he did not have time to look for these works in Paris, and he could not procure them in the provinces.

7 Fresnel to Arago, September 23, 1815, in Fresnel, *Oeuvres* 1:5–6; italics added.

Three weeks later, Fresnel described how he made his discovery in a paper on diffraction, which he sent to the *Académie des Sciences*:

> I have already glued *many times* a small black paper square to one side of an iron wire which served me in my experiments, and I have always seen the internal fringes disappear against this paper; however I only sought its influence on the external fringes and in a way I refused myself the remarkable consequence to which this phenomenon led me. It struck me when I occupied myself with internal fringes, and I have immediately made up this consideration: since by intercepting light from one side of the wire one can make the internal fringes disappear, thus a meeting of rays which arrive from *two* sides is necessary for their production.[8]

Now, Fresnel explains the formation of the fringes:

> They cannot arise from a simple mixture of these rays, since each side of the wire separately casts into the shadow only a continuous light; it is their meeting, the very crossing of these rays which produces the fringes. This consequence, which, is only, so to speak, a *translation of the phenomenon*, totally opposes to the hypothesis of Newton and fully confirms the theory of vibrations. One conceives easily that the vibrations of two rays, which cross at a *very small angle*, can oppose one another when the *node of one corresponds to the loop of the other*. It is this which accounts, without doubt, for the fringes inside the shadow as well as outside it. Outside, the fringes are produced by the intersection of rays departing from the luminous point and from the edges of the wire, and inside the shadow they arise from the intersection of rays inflected by each side of the wire.[9]

There is no question that Fresnel means that two waves destroy one another when an anti-node of one meets an anti-node of the opposite phase of the other wave, rather than when a node meets an anti-node. Fresnel repeated this error in

[8] Fresnel, "First Memoir," 16–17; italics added.
[9] Ibid., 17; italics added.

several papers, which means that he was little familiar with wave phenomena.[10] This is not suprizing for he could not learn much about interference of sound or water waves either from Hassenfratz, his physics professor at the *École Polytechnique*, nor from Haüy's textbook, popular in this school.[11]

Fresnel's line of thought appears very convincing, except for one point: it is not clear why he used the paper squares to study the external fringes. Fresnel explains this as follows:

> I have noticed that when the metal wire was too thin the external fringes became slightly concave against the shadow of the paper, from which I concluded that the light inflected by one side of the wire can sensibly influence the external fringes on the other side; this made me keen to use in my experiments only wires having at least one millimeter in diameter. One cannot suppose that the small paper acts by attraction on the rays which pass on the other side of the wire, since it is too removed from them. Moreover, the fringes do not vary with the mass or the surface of bodies against which light is inflected. The cutting edge and the back of a razor, a metal wire either polished or covered with soot always give the same fringes.[12]

Let us examine the three passages to determine the role of the paper square. First, the experiment with the paper square was not at all an accidental observation: it was repeated "many times." Second, Fresnel's purpose was not to study the effect of the mass, surface, and reflective ability of a diffracting body on the fringes, because he does not vary the attachment and because he says clearly that he knew the paper should *not* affect the inflection of light on either side due to

---

[10] He repeated this mistake twice in his "First Memoir" and retained it in tis published version (see "Second Memoir," 94, footnote by Verdet). Arago copied Fresnel's terms in his report (see "Rapport fait à la Première Classe de l'Institut, le 25 mars 1816, sur un Mémoire relative aux phénomènes de la diffraction de la lumière par M. Fresnel," in Fresnel, *Oeuvres* 1:82, henceforth cited as "1816 Report." I believe that both Fresnel and Arago adopted the terms "nodes" and "antinodes" from textbooks which they studied at the *École Polytechnique* rather than from the original papers of Sauveur, D. Bernoulli, and others. A possible source is Haüy's *Traité de physique* 1:307–8. When the March issue of *Annales* appeared, someone (possibly Biot) indicated the error, which Fresnel corrected in some copies of the journal, replacing "nodes" and "antinodes" with "condensed and dilated nodes"; *Oeuvres* 1:17, footnote by Verdet, and also "Second Memoir," 94. He continued to apply these meaningless terms even later in 1816; *Oeuvres* 1:394 n.

[11] See Hassenfratz, n.59, Ch. III and Haüy, *Traité de physique* 1:304–20.

[12] Fresnel, "First Memoir," 16 n.

[13] Fresnel believed that Young did not attach his screen to the wire in his "screening" experiment to avoid an increase of its mass (Fresnel, "Second Memoir," 93, n.2.).

these causes.[13] Thus, whatever the purpose of the experiment, the attachment was selected so as to *exclude* the effect of its mass, surface, and reflectivity.

What might have been then the purpose of the experiment with the paper square? My conjecture is that it was to stop light, which Fresnel needed to study whether intercepting light on one side of a wire will change the external fringes on the other side of it. Furthemore, I suggest that this experiment was to test the idea of "the influence of rays on one another," because Fresnel was looking for a specific kind of influence. It was not the change of curvature of the fringes formed by a thin wire although it was, in his view, a sign that "the light inflected by one side of the wire can sensibly influence the external fringes on the other side". Nor was it a mixture of two beams which can only change the brightness of fringes. Fresnel was looking for a *disappearance* of fringes, and he immediately recognized the effect when he turned his attention to the internal fringes. Thus his original goal was to prove that the external fringes are formed by the rays inflected by two sides of a wire. He found this to be true for the internal fringes, while the origin of the external ones turned out to be different. What is important to us, however, is that from the very beginning of his experiments (those that we know of) Fresnel was searching purposefully for an intersection of two rays, which destroys light.

I must emphasize here again that although various features of the diffraction fringes, including disappearance of the internal fringes, had been studied for a century and a half before Young, no one ever considered them to be a two-ray effect. This view was introduced by Young and, as shown in the next chapter, it was widely accepted after 1815. Since this approach was not a "natural" one and since there is no evidence that Fresnel tried other hypotheses until he found empirically this one, I would conjecture that he utilized Young's idea, which came to him through Arago, or from Young's *Lectures*.

It took Young about two years of intensive experimental and theoretical studies to discover the principle of superposition of waves and temporal interference of sound and two more years to pass from those to interference of light. This fact and the very slow reception of Young's ideas shown above, demonstrates that the concept of interference was difficult to discover and to grasp. I believe that even for a genius, such as Fresnel, to discover the principle of interference of light in less than five months without any previous experience in wave phenomena and without any hint from outside, would be a miracle.

Verdet, referring to the note which Fresnel received from Arago, says that his advices "have not been of great direct usefulness for the young physicist," because Fresnel was unable to read in English.[14] This ignores, however, a possibility

[14] Verdet, "Introduction," XXX.

for Fresnel to benefit from his talks with Arago. There are several places in Fresnel's early papers where he acknowledges an information obtained from Arago. Since these papers were written before Fresnel's next visit to Paris (January or February 1816), and there is no indication that they came in letters while there are references to things *spoken*, I believe that Fresnel received this information from Arago in July 1815.

Among the subjects of their conversations were: 1) a possibility of verifying whether light travels faster in air than in water by means of a water-filled telescope ("you *told* me of the experiment . . ." [15]); 2) Arago's hypothesis that a small change of the velocity of light particles does not affect their refraction;[16] 3) forming a luminous point by means of a very convex lens: "the means which M. Arago has indicated to me";[17] and 4) the colors of striated surfaces: "I had never seen this phenomenon and I was not sure that I remember well what you have *told* me about it."[18]

The latter phenomenon might have been referred to in connection with Young's 1801 experiment because his work was a subject of discussion. While explaining to Arago why he was unable to benefit from his note Fresnel says among other things:

> The Philosophical Transactions are, I think, a periodical work which I can consult only in Paris. As to the work ("ouvrage") of Young, *of which you have told me so much*, I had a strong desire to read it, but not knowing English, I could only listen to it with the help of my brother, and after leaving him, the *book* became *again* unintelligible to me.[19]

Young's work, of which Arago told Fresnel "so much," is probably his *Lectures*, because it is not in the *Phil. Trans.* and is referred to as a "book". The fact that the *Lectures* "became *again* unintelligible" means that a part of it *was* read with the help of Fresnel's brother Fulgence before they parted.

It is naturally to suppose that the first thing that would have interested Fresnel in the *Lectures* was the nature of light. If Fulgence translated for him, the chapter "On the nature of light and colours," Augustin would have been able to grasp the idea of the combination of two rays of light. If he did not get it from the *Lectures*, he could obtain it from Arago. It is difficult to imagine that Arago, who was so

---

15 See n.7 above, 7; italics added.
16 Fresnel, "First Memoir," 11.
17 Ibid., 14. The discussion of this technical detail suggests, in my view, that Fresnel did some experiments on diffraction prior to his visit to Paris.
18 Fresnel to Arago, November 20, 1815: Fresnel, *Oeuvres* 1: 68; italics added. See also his "First Memoir," 30.
19 See n.7 above, 7; italics added.

well familiar with Young's papers that he immediately recognized Fresnel's method as a copy of Young's (see VII.2), did not mention *anything* about the principle of interference. He could have done it while discussing either Young's *Lectures* or his experiment with the colors of a striated surface.

It is strange that Fresnel did not acknowledge his debt to Young. A possible explanation is that *after* he had tried the hypothesis that two light rays affect one another, Fresnel became convinced that it is a *trivial* idea for waves and thus it cannot be someone's invention. (As shown above, Young also believed that the principle of interference of waves was a common knowledge.) What mattered to Fresnel was a proper application of this idea; and since he learned from Young no equations, no experimental details, no mathematical laws of phenomena, he did not believe that he borrowed anything. Actually, even if he knew the description of Young's approach to diffraction on a narrow body in the *Lectures*, Fresnel borrowed from it only the idea of two interfering rays but not how to choose them.[20] It is worth noting how Fresnel tried to clarify the question of priority in his letter to Young: "When I submitted it [the "First Memoir"] to the Institute, I did not know your *experiences and the consequences* which you had drawn from them."[21] There is no word here about the principle of interference ("consequences," refer, in my view, to such results as the curvilinear propagation of fringes and the like). This supports my suggestion that Fresnel did not view this principle as a discovery to be credited for.

Thus, my conjecture is that Fresnel learned the idea that light rays can affect one another either from the conversation with Arago or from Young's *Lectures*. His diffraction experiments (at least a considerable part of them, for he also studied the effect of mass and surface) were directed to test this hypothesis. He did it differently from Young, in particular, he proved it initially for the internal fringes.

Thus, while other scientists could not understand even a detailed presentation of Young's principle of interference of light, Fresnel needed only a hint to rediscover it. Fresnel, of course, was a genius and well prepared for his task, but it seems that his success was to a considerable degree due to his great interest in proving the wave theory. Others did not have a sufficient motivation for doing extra work on Young's papers: guessing the meaning of his statements, checking his mathematics, and verifying his experiments.

It is important to note that in the beginning Fresnel considered the principle of interference to be a "translation of phenomena", fully independent of the wave

[20] According to Young, the external fringes are produced by interference of a direct and a reflected rays (*Lectures* 1: 467), while Fresnel apparently began with the hypothesis that these fringes result from an intersection of rays passing by two sides of a wire.

[21] Fresnel to Young, May 24, 1816: Fresnel, *Oeuvres* 2: 737–38; italics added.

theory, since he was frequently speaking of his theory as of "theory of vibrations *and* of the infuence of rays on one another."[22] That was his natural attitude, for that was before the beginning of the promotional campaign launched by Arago. This suggests that perhaps Young's tendency to do the same was, at least, in the beginning, also a natural rather than a tactical move. Indeed, Young's *Syllabus*, in which he carefully separated facts from theories, was written before his "Theory of light" appeared in print, and consequently he could not know that his version of the wave theory will not be accepted.

Since Fresnel viewed the principle of interference only as a means for studying diffraction, let us see how he did it.

b) Diffraction by a narrow body

As shown above (n.9), he selected the interfering rays in the same way as Young. In his "First Memoir" he presented (without derivation) the equation for the first external minimum

$$x = \sqrt{\frac{\lambda b(a+b)}{a}}, \tag{7.1}$$

where x is the distance of the fringe from the shadow, a and b are the distances of the body from the light source and the screen, respectively, and λ is the wavelength. He assumed the "reflected" ray had its phase inverted.[23] In the published version of the paper he gave the derivation of eq. (7.1). He also derived the breadth of the internal fringes

$$\delta = \frac{\lambda b}{d}, \tag{7.2}$$

where d is the body's breadth.[24]

---

[22] Fresnel, "First Memoir," 33; Fresnel to Arago, October 26, 1815: Fresnel, *Oeuvres* 1: 35, 26; and "Second Memoir," 117, italics added.

[23] Fresnel, "First Memoir," 18, 26.

[24] Fresnel, "Second Memoir," 97, 106–7.

When measuring the external fringes Fresnel took the border between the red of the first order and the violet of the second order to be the first dark fringe. He also supposed that this fringe corresponds in Newton's table of the thicknesses of thin films reflecting particular colors to a sum of the thicknesses of air which reflect the two colors.[25] In this way he obtained the wavelength 517.6 nm, which is 10% less than that which, according to Newton's measurements, corresponds to the brightest part of the spectrum (588.9 nm).

Let us now evaluate the precision of Fresnel's early measurements of the external fringes. For this purpose he used the difference between the predicted and measured position of a fringe $x_{exp} - x_{th}$; in his observations with white light this was about 0.1 mm in 1815 and 0.03 mm in 1816.[26] The average relative error $(x_{exp} - x_{th})/x_{th}$, which is a better criterion, would be for the same observations respectively 3.5% and 2.5%. This accuracy seems to be about four times better than in Young's 1803 paper (see Table 5). However, it is easy to see that Fresnel's measurements were in fact no more precise than Young's. Let us calculate the wavelengths from Fresnel's measurements of the positions of the fringes, as Young did. Let λ be the average wavelength, s the standard deviation, and s/λ the relative error. Then, the precision of Fresnel's observations of the external fringes in white light is 6.8%–7.5%. Now, after Young introduced the phase inversion, the discrepancy between his own observations and the theory decreased to 4.7% (Table 5). Unfortunately, he did not publish his corrected wavelengths. Thus, deliberately or not, Fresnel chose a form for the comparison of theoretical and experimental results that diminished the discrepancy between them. Since Young believed that a difference of 10–15% is acceptable, he did not see any need in publishing the corrected results that he had already obtained. Nor did he intend to exclude Newton's Observation 9, which evidently was less exact than the other experiments presented by Young.

---

[25] This sum is equal 20.167×10^{-6} in (see Newton's *Opticks*, 233) or 512.2 nm. Fresnel was probably not aware that the English inch (25.4 mm) differed from the Parisian inch (25.66 mm) and obtained 517,6 nm instead of 512.2 nm. Fresnel's rule may be clarified in the following way. The path difference for the first minimum in reflected light ist $2e=\lambda$. Let e_r and e_v be the thicknesses which reflect the red of the first and the violet of the second order, respectively. If $e=(e_r+e_v)/2$, we have $\lambda=e_r+e_v$, as Fresnel claimed. Since the maxima of the first two orders correspond to the equations $2e_r=\lambda_r/2$ and $2e_y=3\lambda_v/4$, we obtain $e_r+e_v=(\lambda_r+3\lambda_v)/4$. By substituting $\lambda_r=914$ nm and $\lambda_v=378$ nm (see Table 1) we get $\lambda=512$ nm, as Fresnel ought to have obtained.

[26] Fresnel, "First Memoir," 15, 19, 20, and "Second Memoir," 99, 101. For another method of evaluating Fresnel's precision see Buchwald, *Wave Theory*, 123–127.

c) Reflection and refraction of light

In addition to the diffraction of light by a narrow body, Fresnel successfully applied the principle of interference to prove the laws of reflection and refraction of light. Let us consider his reasoning in the case of reflection of light. Let *AB* be a reflecting surface (Fig. 33) and *EG* and *FD* two parallel incident rays. These

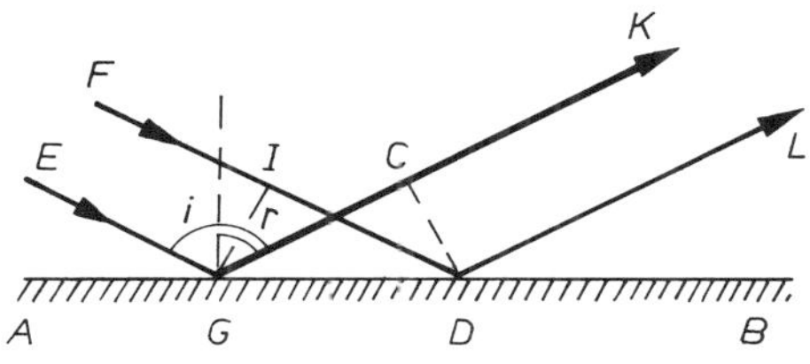

Fig. 33
Fresnel's theory of reflection (from *Oeuvres* 1: 29)

rays are coherent at the points *E* and *F* (as Fresnel says, they "vibrate in accord"), where *GI* is normal to the incident rays. Fresnel assumes that reflection takes place in all possible directions. Then, he says, if $GC \neq ID$, or if the angle of reflection *r* differs from the angle of incidence *i*, the points *C* and *D* at the normal *CD* to the reflected rays *GK* and *DL* will no longer vibrate in accord, and one can find such a distance *GD* that the path difference $P = \lambda/2$, and the reflected rays will destroy one another (λ is the wavelength). In this argument the path difference of reflected rays is

$$P = a(\sin r - \sin i), \tag{7.3}$$

where *a* is the distance between the points of reflection. Thus, Fresnel may be understood as saying that given the angle of incidence, for any angle of reflection there exists a distance

$$a = \frac{\lambda}{2(\sin r - \sin i)}, \tag{7.4}$$

such that the light reflected in this direction is destroyed. Thus, the reflected light will be perceived only at the angle $r = i$, for at this angle the vibrations in both rays will always be in phase. Fresnel indicated in his "First Memoir" that this reasoning is valid for a real surface provided that its irregularities are consider-

[27] Fresnel, "First Memoir," 29.

ably smaller than the wavelength. The idea that every point of a surface can be considered as a center of spherical waves later helped Fresnel to formulate the Huygens-Fresnel principle. None of Fresnel's contemporaries, however, recognized the importance of his account of reflection and refraction, at least before 1819.

d) Other applications

To make a convincing case for the principle of interference, a month after submitting the "First Memoir" Fresnel added a supplement to it which contained applications of the principle to other phenomena. One of them was for the colors of thin films, the account of which was very similar to Young's. In particular, Fresnel's equation for the path difference is identical to eq. (5.3), since he also introduced the inversion of phase to account for the central black spot. Although unaware of it, Fresnel followed Young step by step, with the only difference being that he called attention to the discrepancy between the theory and Newton's observations at very large angles of refraction and claimed that Newton was wrong.[28] Fresnel himself never studied this phenomenon experimentally.

Another experiment that Fresnel did not perform and discussed only theoretically was the colors of striated surfaces, which he learned about from Arago. The underlying idea here was the same as in his theory of reflection and refraction, namely, that particles of the surface become new centers of spherical waves, although now the particles were limited to those lying on the scratches *A* and *B* (Fig.34).

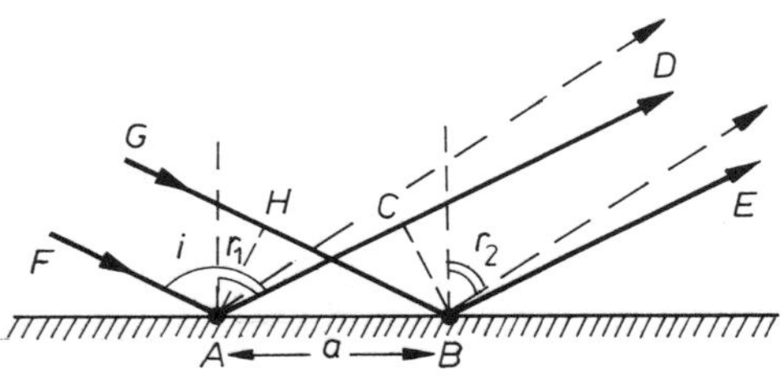

Fig. 34
Fresnel's theory of diffraction on a grating (from *Oeuvres* 1: 45)

[28] Fresnel, "First Supplement," 51–58.

According to Fresnel, the path difference of the interfering rays will be where a is

$$P = a(\sin i - \sin r), \tag{7.5}$$

the distance between two scratches, and i and r are the angles of incidence and reflection, respectively. The bright fringes will be seen at an angle r such that

$$\sin r = \sin i - \frac{m\lambda}{a}, \tag{7.6}$$

where λ is the wavelength, and m an integer. This equation is more general than Young's eq. (5.16), since Fresnel considered the angle of incidence as a variable. He realized that such a grating can under certain conditions separate the colors of white light. Assuming that light of wavelengths λ_1 and λ_2 is diffracted at the angles r_1 and r_2 respectively, Fresnel obtained (for small r)

$$r_2 - r_1 = \frac{m(\lambda_2 - \lambda_1)}{a \cos r}. \tag{7.7}$$

The value $a \cos r$ may be considered as an "effective distance a," i.e. inclining the grating to the incident light is equivalent to shortening the distance between the scratches. To test this theory, Fresnel used two threads. Considering the diffraction in transmitted light, he derived the path difference as

$$P = a(\sin i + \sin r). \tag{7.8}$$

At normal incidence the bright fringes are found at

$$\sin r_{\max} = \frac{m\lambda}{a}, \tag{7.9}$$

while for the dark fringes we have

$$\sin r_{\min} = \frac{(2m-1)\lambda}{2a}. \tag{7.10}$$

To verify eq. (7.10) Fresnel observed the fringes around two threads making a small angle with each other. In three of his four observations the relative errors ranged from 6% to 12%, and he considered this sufficiently accurate.[29] The equations (7.6)–(7.9) for the maxima are correct, although Fresnel's derivation is invalid, for he considers only two interfering rays for a grating with a large number of lines, exactly as Young did in 1801. However, equation (7.10) for the minima is

29 Fresnel, "First Supplement," 45–51.

erroneous.[30] This error, though, was not known to readers of the published paper, for when Fresnel was informed by Arago that Young had already examined the colors of thin films and those of striated surfaces, he deleted the discussion of these phenomena from the published paper.

In the same volume of the *Annales de Chimie et Physique* where Arago published Fresnel's 1816 memoir on diffraction, Arago also briefly reported on two new phenomena of interference which he observed together with Fresnel. One of them was a version of Young's "screening" experiment.[31] They found that the fringes within the shadow of a body vanished even when the screen intercepting light from one side of the body was a transparent plate. Since the plate did not stop the light, this phenomenon apparently contradicted the claim that *two* beams of light are necessary to produce fringes. Fresnel suggested that the fringes did not vanish but simply shifted out of the shadow and became invisible in the bright direct light. He explained that the shift was due to a delay of the ray passing through the plate relative to the other ray traveling in air. He suggested to Arago that a very thin piece of glass or mica be used to get a small fringe shift that would be observable within the shadow. If light travels slower in a denser medium, the fringes must move toward the plate. The experiment fully confirmed Fresnel's predictions.[32] Placing two plates of glass of different thickness at the opposite edges of the body, Arago was able to shift the fringes in either direction. This phenomenon, together with the colors of thin films and those of "mixed plates," showed that the principle of interference *requires* the assumption that light propagates slower in a denser medium.

The other observation reported by Arago was the famous experiment with "Fresnel's mirrors."[33] Fresnel and Arago observed the fringes produced by two mirrors making an angle of almost 180° with each other. The idea for the experiment and the theory underlying it were Fresnel's; Arago suggested the use of metal mirrors instead of glass ones and showed that intercepting one of the two beams by a transparent plate made the fringes disappear.[34] Fresnel and Arago believed that this experiment refuted the view of the "emissionists" that fringes

[30] Verdet believed that despite the erroneous derivation, Fresnel's formula (7.10) for the minima can represent diffracton by two threads (Fresnel, *Oeuvres* 1: 49 n). The origin of his conclusion is unclear, since no one theoretically studied such a case. He possibly, believed that it is identical to the diffraction by two slits, because two slits give the same position for the minima as a grating. However, it is easy to show that the phenomena are not equivalent.

[31] Arago, "Note sur un phénomène remarquable qui s'observe dans la diffraction de la lumière," *Oeuvres de Fresnel* 1: 75–77

[32] A. Fresnel to Leonor Fresnel, March 4, 1816, *Oeuvres* 2: 834.

[33] Arago, "Remarques sur l'influence mutuelle de deux faisceaux lumineux . . .," *Oeuvres de Fresnel* 1: 123–5.

[34] Fresnel, "Supplément au deuxième mémoire sur la diffraction de la lumière," *Oeuvres* 1: 152 (this is unpublished and henceforth referred to as the "Second Supplement").

are formed by the *edges* of bodies acting upon light. This experiment also provided Fresnel with a new method to measure wavelengths that afterward played an important role in improving the precision of the diffraction observations in his prize-winning memoir.

e) Early formulations of the principle of interference and the conditions of coherence

Despite Fresnel's tendency to a detailed exposition, the physical meaning of the principle of interference does not emerge from his papers more clearly than from Young's. First, Fresnel does not give any clear definition of the principle of interference. His original formulation F(1) (see VII.1) is worse than any of Young's: it contains an error and lacks important details, in particular, it mentions only one condition of coherence. In the published version of this paper, the error is "corrected" with another error, the rest remaining unchanged.[35] Fortunately for French readers, Arago provided them with better definitions of the principle of interference, which he copied from Young. For instance, in March 1816, Arago stated:

A (1) 1816 — One owes to Doctor Thomas Young for . . . demonstrating the first, by the experiment with internal diffraction fringes . . ., that two homogeneous rays from the same origin, which arrive at a point by two slightly different routes, can destroy themselves, or, at least, considerably weaken one another.[36]

He gave another version of it in his article on scintillations of stars, which, however, hardly reached many scientists interested in optics:

A (2) 1816 — According to the experiments of this celebrated physicist [T. Young] two rays of homogeneous light, which arrive at a point in space by two slightly unequal routes, reinforce or destroy themselves according to the difference of the paths they travelled being of this or that magnitude.[37]

Second, Fresnel was not very clear about the conditions of coherence. Like Young, Fresnel was concerned with the problem of coherence from the very be-

[35] See n. 10

[36] See n. 33, 123–24.

[37] Arago, *Oeuvres* 7:98. See n. 10, Ch. VIII.

ginning of his work on interference. His term for coherence was the "accord of vibrations." In contrast to Young, he explicitly raised the problem of why in some cases the vibrations are coherent whereas in others they are not. In the early period Fresnel devoted most attention to the condition of a common source, and I will begin with his views on this subject.

Fresnel realized that one of the most powerful objections against the principle of interference could be the non-interference of light from two different sources. He originally turned to this problem in his "First Memoir." Apparently, he received his first clue for the necessity of a single source of interfering light from observations. Initially, he measured the paths of rays forming the internal fringes from the edges of a body. Then he discovered that when a plate was inclined to the incident light, it did not alter the symmetry of the fringes about the central white fringe. Nor did he find that the fringes' location depend on the surface irregularities of a cylinder. Both observations contradicted his hypothesis, and to save the phenomena Fresnel supposed that the paths of interfering rays must be measured from the luminous source and not from the diffractor.[38] A month later Fresnel explained why light from two sources cannot interfere. His principal idea was that interference fringes are visible only if they are *permanent*, and this implied that the phase difference of interfering rays must be constant. Fresnel realized that the non-interference of light from different sources means that light waves are discontinuous, or that "the agreements and disagreements of light rays coming from different sources vary every instant, and because of that they cannot produce a *constant and consequently perceptible* effect."[39] In the published version of the memoir he suggested that particles of bodies vibrate independently, and therefore,

> if there is no dependence between the centers of vibration, the *moment* of the departure of a wave system is not connected with the moment of the departure of neighboring waves, since any cause that engenders them does not produce simultaneous changes in the two luminous sources; that is why the lines of agreement and disagreement will continuously vary their positions, and the eye will get only a sensation of uniform light; it is this that undoubtedly impeded us for so long from recognizing the influence that light rays exert on one another.[40]

Two points in this passage deserve a comment. First, Fresnel called attention to the uncommon character of the phenomena of interference, which require spe-

[38] Fresnel, "First Memoir," 26.
[39] Fresnel, "First Supplement," 43. Italics added.
[40] Fresnel, "Second Memoir," 94.

cial conditions to be observed. Second, he assumed as early as 1816 that light waves are of *finite length*. Bearing in mind that he insisted from the very beginning on the periodicity of light waves, we must acknowledge that at the time Fresnel already used, though only implicitly, the concept of wave trains.

After formulating the necessity of a common source, Fresnel realized that instead of a single primary source one can sometimes use two secondary sources, such as the edges of a narrow body. Originally, he treated these sources simply as mathematical centers of waves, but soon he discovered a physical reason for these centers to be two *coherent* oscillators. According to Fresnel, the incident light makes particles at the edges vibrate, and consequently these particles emit light of the same frequency as the primary one.[41] Later he also considered such secondary sources as two slits or virtual images of a luminous point in "Fresnel's mirrors." Besides secondary sources, Fresnel discussed artificial sources such as a small hole in a shutter or the focus of a lens collecting sunlight. The question of why these artificial sources may be treated as luminous points first appeared in the "First Memoir." In the next two months Fresnel presented a solution.

When dealing with a hole he distinguishes between the fringes produced by light diffracted by the border of the hole and those formed by undiffracted, or "direct" light. He claims that the fringes produced by direct light do not disturb those formed by diffracted light, but his demonstration is too obscure to be discussed here. On the other hand, his conclusion about interference of diffracted light is important to us. Fresnel says that since the diffracted waves have their centers at different points of the hole's border,

> the accord of their vibrations does not extend indefinitely far from the axis of the light pencil. But the space in which they noticeably agree is in inverse ratio to the breadth of the hole and becomes considerable when the hole is sufficiently narrow.[42]

This reminds us of a statement from a modern optics textbook that the larger the source, the smaller that part of space where light rays are coherent, and vice versa. Fresnel seems to be saying that when the angle of diffraction increases, the overlapping of fringes produced by different centers increases too, and that the

[41] "Fresnel to Arago, December 3, 1815," *Oeuvres* 1:72 and "Second Memoir," 113.
[42] Fresnel, "First Supplement," 43.

larger the hole the more significant the effect. It is possible that Fresnel's reasoning was influenced by his study of the external fringes.[43]

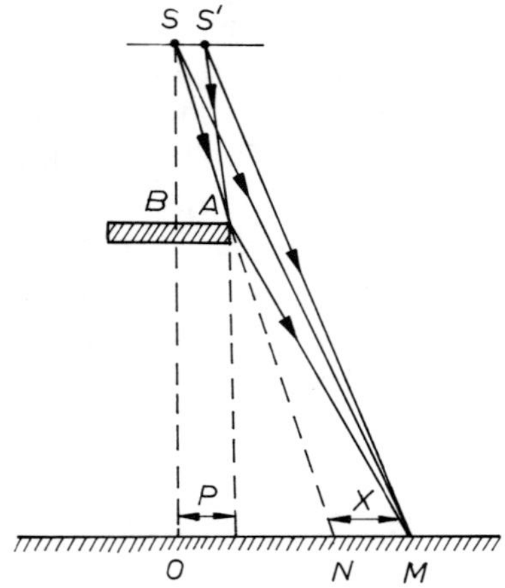

Fig. 35
An illustration to Fresnel's consideration of the size of a hole as a secondary luminous source

[43] Let S be the center of the hole of radius r, and S' a point at its border (Fig. 35). These two sources create at M two fringes which correspond to the path difference $P = SA + AM - SM$ and $P' = S'A + AM - S'M$, respectively. These fringes coincide when $P' - P$ is small compared to the wavelength λ. Let p, r and x be small relative to the distances a and b from the hole to the source, and observation screen, respectively, $p = AB$, $x = MN$. Then $P' - P = (S'A - SA) + (SM - S'M)$, and

$$SA = \sqrt{a^2 + p^2}\,; \quad S'A = \sqrt{a^2 + (p-r)^2}\,;$$

$$SM = \sqrt{(a+b)^2 + \left[\left(\frac{a+b}{a}\right)p + c\right]^2}\,;$$

$$S'M = \sqrt{(a+b)^2 + \left[\left(\frac{a+b}{a}\right)p + x - r\right]^2}\,;$$

so that we obtain

$$P' - P = \left(xr + \frac{br^2}{2a}\right) / (a+b).$$

Thus, when the hole is large, $P' - P$ may reach $\lambda/2$ even for the fringes of the first order (small x), thus such a shift of fringes would obliterate all fringes. On the contrary, when the hole is small, even the fringes of higher orders (large x) are distinct.

In the case of a lens, Fresnel's considerations are more lucid. He had a twofold practical interest in this investigation. First, he was looking for the best point source of light. Second, he needed a proof that with the direct method of observation of interference, the fringes exhibited at the retina are of exactly the same size as those seen on a screen. Fresnel was searching for the conditions under which refraction in a converging lens would least affect the path difference of interfering rays. Let two parallel rays DA and EB (Fig. 36) be refracted by the lens (O is

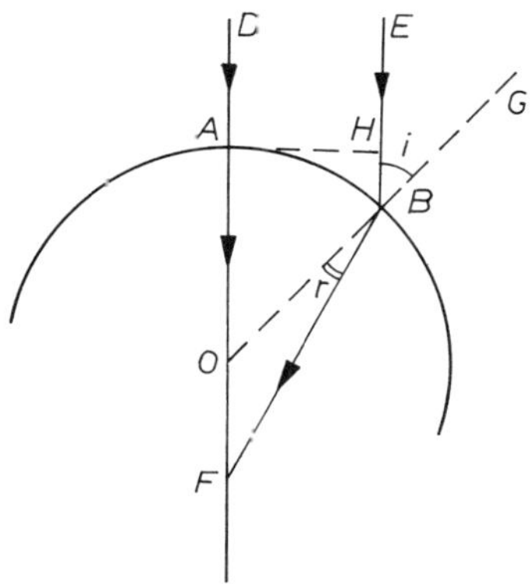

Fig. 36
Fresnel's consideration of the role of a short-focus lens as a secondary luminous source (from *Oeuvres* 1: 44)

the center) and intersect at F. Their path difference is $P = BH + nBF - nAF$, where n is the index of refraction. For a small angle of incidence i, Fresnel obtained the path difference

$$P = \frac{r(n-1)\sin^4 i}{8n^3}, \qquad (7.11)$$

where r is the lens's radius. He calculated from this equation that for a lens 1 cm in diameter, a beam of coherent light with an angular divergence of about 11° acquires after refraction an additional path difference of about one tenth of the wavelength.[44] In other words, such a lens does not change coherence of the incident beam, and its focus F may serve as a luminous point.

[44] Fresnel, "First Supplement," 44–45.

As to the other conditions of coherence, Fresnel was rather terse. Apparently, he based the necessity for interfering rays to be almost of the same direction on the relation

$$\delta = \frac{\lambda}{\alpha}, \tag{7.12}$$

where α is the angle between the rays, λ is the wavelength, and δ is the fringes' breadth. When this angle increases sufficiently, the fringes become too narrow to be seen. On the contrary, when the angle approaches zero, the fringes are too broad to be perceived. This means that the whole field of vision is covered by one fringe, and this uniform illumination would lead us to conclude that there are no fringes at all. Though Fresnel did not explicitly present eq. (7.12) before July 1816, some of his remarks indicate that he had already possessed it earlier (I assume $\alpha \simeq \sin\alpha \simeq \tan\alpha$, for he considered only very small angles α). For instance, in October 1815 he said that, "as we move away, the undulations cross at a smaller angle, and the fringe broadens."[45] In March 1816 Arago reported that the width of the fringes produced by "Fresnel's mirrors" is inversely proportional to the distance d between the secondary sources, which is obviously a part of the formula $\delta = \lambda b/d$, where b is the distance between the secondary sources and the observation screen.[46] By the trivial substitution $d/b \simeq \alpha$, the latter formula may be transformed into eq. (7.12).

In his "First Memoir" Fresnel raised the question of why rays of different colors do not destroy each other even when they come from the same source in almost the same direction. He suggested two explanations. First, the waves of different frequency do not arrive at the eye simultaneously.[47] Second, if they do arrive at the same time, they never oppose each other as completely and permanently as vibrations of the same frequency and opposite phase do.[48] He is probably hinting here at the explanation, which he published later, that such an interference is similar to beats of sound but occurs too rapidly to be recognized. This passage was not included in the published version of the paper.

Fresnel, like Young, followed Newton in explaining the limit on the number of fringes seen in thin films illuminated by white light. He remarked in his "First Supplement" that by utilizing monochromatic light one can observe dark fringes even where the thickness of the air film is considerable.[49] He also mentioned in

---

[45] Fresnel, "First Memoir," 21.

[46] Arago, "Influence mutuelle de faisceaux lumineux," 124.

[47] Fresnel, "First Memoir," 27. Since Young also held to the sequential model of light (see his "Theory of light," 148–49), he might have had a view similar to Fresnel's.

[48] Ibid.

[49] Fresnel, "First Supplement," 52 n.

the published paper that while only three external fringes can be seen in white light, eight are visible in monochromatic light.[50] Fresnel believed that the farther the fringes from the border of the shadow, or, in other words, the higher their order, the greater their overlapping. He erroneously supposed another cause limiting the number of fringes, namely, the reduction of intensity of the fringes of higher orders due to their larger distance from the diffracting body.[51] During this period, though, Fresnel made no general statement about the limitation of the path difference, nor did he give any numerical example.

Let me now summarize Fresnel's work as presented in his 1816 memoir. His major achievements consisted of a clear graphical illustration of the principle of interference, and the development of mathematical theory of diffraction and its experimental verification. As will be shown in the next chapter, this part of his work was immediately acclaimed by several French scientists. The situation was much different with the physical part of Fresnel' theory. Although Fresnel's explanations were much more detailed than Young's, most of them were quite obscure. His formulation of the principle of interference was vague and contained an error. Of all the conditions of coherence only that of a common source was explicitly formulated and partly explained. Fresnel's readers hardly had a better opportunity to comprehend the physical meaning of the principle of interference than Young's ones.

VII.2 Second period (1816–1818)

In July 1816, Fresnel attempted to improve his explanation of the internal fringes by taking into consideration an infinite number of interfering waves, which originate at different parts of the incident wave.[52] Since at the time he could add vibrations only of the same or of the opposite phase, Fresnel had to reduce his theory again to two-ray interference, though now the interfering rays originated at a small distance from the edges of a diffractor. In the fall of 1817, while working on chromatic polarization he solved the problem of the addition of more than

[50] Fresnel, "Second Memoir," 103.
[51] Ibid, 96.
[52] Fresnel, "Second Supplement," 161–64.

two waves with arbitrary phases.[53] Fresnel considered harmonic waves of light of the type

$$a \cos\left[2\pi\left(t - \frac{x}{\lambda}\right)\right], \tag{7.13}$$

where a is the maximum velocity of the ether molecule, x its position, t the time, the unit of which is the period, and λ the wavelength.[54] He began with an additon of two waves that have the same wavelength λ, different amplitudes a_1 and a_2, and a phase difference of 90°, and found it to resemble an addition of two perpendicular forces. The resulting amplitude was

$$A = \sqrt{a_1^2 + a_2^2}\,. \tag{7.14}$$

In the case of many interfering waves, Fresnel resolved each of them into two perpendicular components, separately added the components and reduced the final result to the form given by eq. (7.14).

Fresnel applied this idea in his 1818 theory of diffraction. This theory was based on the Huygens-Fresnel principle, which stated that,

> The vibrations at each point of a light wave may be treated as the sum of elementary movements which are sent there, independently of one another, by all the parts of the wave, considered in one of its previous positions.[55]

[53] A. Fresnel to L. Fresnel, November 28, 1817, *Oeuvres* 2:844.

[54] Fresnel, "Supplément au mémoire sur les modifications que la réflexion imprime a la lumière polarisée [1818]," *Ouevres* 1:490. Fresnel assumed the period of vibrations T = 1.

[55] Fresnel, "Mémoire couronné," 293.

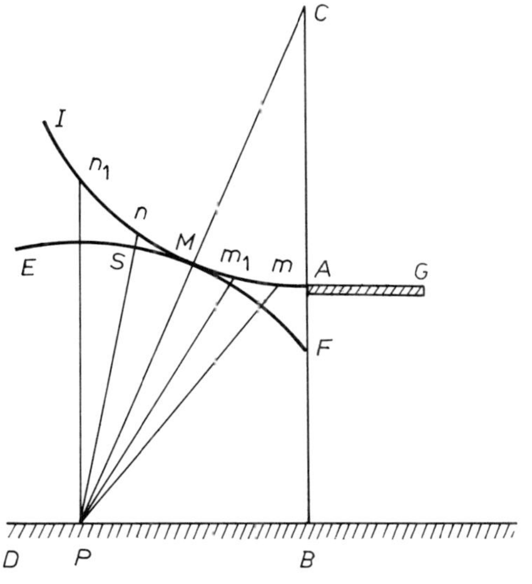

Fig. 37
An illustration of Huygens-Fresnel principle (from *Oeuvres* 1: 313)

This may be illustrated in Figure 37 where an incident wave AMI is partly intercepted by a screen AG. To evaluate the intensity of light at point P of the observation screen DB, Fresnel divided the wave AMI into infinitely small arcs Am, mm_1, etc, and then added the waves coming from these arcs to the point P. By assuming that the secondary waves have a significant amplitude only in directions nearly parallel to that of the primary wave, he deduced that the components of the secondary wave, which is emitted by an arc of length dz and distant from M by z, are

$$dz \cos\left(\frac{\pi z^2(a+b)}{ab\lambda}\right) \text{ and } dz \sin\left(\frac{\pi z^2(a+b)}{ab\lambda}\right), \tag{7.15}$$

where $a = CA$, $b = AB$, and λ is the wavelength. Thus the resultant vibration at P has the amplitude[56]

$$\sqrt{\left[\int dz \cos\left(\frac{\pi z^2(a+b)}{ab\lambda}\right)\right]^2 + \left[\int dz \sin\left(\frac{\pi z^2(a+b)}{ab\lambda}\right)\right]^2}. \tag{7.16}$$

[56] Ibid., 315–6. See also Buchwald, "Fresnel," 81–91.

Fresnel successfully applied this theory to diffraction by a half-plane, a narrow body, and a narrow aperture. He also derived with its help the laws of reflection and refraction. In fact, the role of the Huygens-Fresnel principle was that of providing a general and uniform procedure for the selection of interfering waves, in contrast to Young's rules, which varied from one phenomenon to another. While mathematicians appreciated the generality of the new principle, there is no evidence, that anyone had earlier refused to accept Young's rules because they seemed too specific or *ad hoc*.

Although at this period coherence was not Fresnel's major problem, he did make several important remarks upon it. He presented an explanation of the condition of direction in his account of "Fresnel's mirrors" in July 1816, although it was published only in 1822.[57] In October 1816, in his paper on interference of polarized light, he mentioned that the rays reflected from "Fresnel's mirrors" when entering the eye must have a path difference of no more than six or seven wavelengths.[58] This was Fresnel's first explicit statement on the condition of the path difference, but it was deleted from the published version. It is worth noting that, unlike Young, Fresnel did not include the conditions of coherence into formulations of the principle of interference.

In his prize-winning memoir on diffraction ("Memoire couronné") Fresnel slightly improved his original formulation F (1)

F (2) 1818 When two systems of light waves tend to produce completely opposite movements at the same point of space, they must weaken one another, and even totally destroy if their impulses are equal; and, on the contrary, the oscillations must add when they are executed in the same direction. The intensity of the light will therefore depend on the respective positions of the two wave systems, or . . . of the difference of the paths traversed when they are emanated by a common source.[59]

A complete absence of conditions of coherence is discouraging: some can be found in footnotes, others are missing. The best for this period formulation of the principle of interference and of the conditions of coherence appeared in a

[57] Fresnel, "Second Supplement," 161–64.

[58] Fresnel, "Mémoire sur l'influence de la polarisation dans l'action que les rayons lumineux exercent les uns sur les autres, [October 6, 1816]," *Oeuvres* 1:411.

[59] Fresnel, "Memoire couronnee," 258.

joint article by Arago and Fresnel. Its similarity to A(1) and A(2) reveals Arago's hand:

A/F 1819 1. *Two* rays *of homogeneous* light, emitted from the *same* source, which arrive at a certain point of space by two *different and slightly unequal* paths, reinforce or destroy one another, according to the path difference is being this or that magnitude.[60]

The next two paragraphs specify that the rays reinforce or destroy one another depending on whether their path difference equals an even or an odd number of a certain constant *d*, which is different for light of different colors, such as, etc.

The most interesting thing about this definition is that the authors attribute it to Young. Indeed, it reproduces the principal features of Young's formulation Y(5), except for the condition of direction which is replaced with the condition of path difference. This supports my suggestion that the two conditions are equivalent.

In 1816, Arago and Fresnel discovered a new condition of coherence for polarized light: two rays of light polarized in perpendicular planes and subsequently reduced to the same plane of polarization interfere if they originate from a polarized ray; and they do not interfere, it they originate from natural light.[61] A search for the reason for this non-interference eventually led Fresnel to the discovery of the transverse nature of light waves. Afterward, he showed that this new condition of coherence is a consequence of the condition of a common source, if it is assumed that each luminous particle simultaneously emits many *independent* wave trains polarized in all possible planes which are perpendicular to the direction of light propagation. Only those waves can, therefore, interfere, which are obtained from the same wave train.

While preparing his memoir for the contest on diffraction, Fresnel improved his understanding of coherence. In an unpublished note on diffraction for 1818 he investigated the limitation of the number of visible fringes. According to Fresnel, fringes of high orders become invisible for two reasons: 1) maxima and minima of many different wavelengths occur at the same point, and this produces a uniform illumination; and 2) the higher the order of fringes, the less will be the contrast between bright and dark fringes.[62] In 1819, in the abridged version of his prize-winning memoir Fresnel gave his first explanation of the non-interference

---

60 Arago & Fresnel, "Mémoire sur l'action que les rayons lumineux exercent les uns sur les autres," *Annales de Chimie* 10 (1819): 288, see *Oeuvres de Fresnel* 1: 509. Italics added.

61 Ibid, 521.

62 Fresnel, "Note sur les phénomènes de la diffraction dans la lumière blanche," *Oeuvres* 1: 193–94.

of light of different frequency. He compared it to beats of sound and concluded that the frequency of "beats of light" is too high for interference to be perceived.[63] In the same paper Fresnel revealed for the first time his interest in the role of the size of a primary light source, which he distinguished from that of a secondary source. He stated that the overlapping of diffraction fringes increases as the source becomes larger, and that even a source that is small enough to produce sharp fringes of the first order forms very blurry fringes of higher orders.[64] In the 1819 memoir on diffraction, Fresnel explicitly presented for the first time his model of light: light is a periodical succession of discontinuous waves (he did not use the term "wave train"). Periodicity is necessary, according to him, because the eye cannot sense the effect produced by a single wave train, and the representation of light waves by eq. (6.13) relates only to the regular part of a wave train, which could be considered as infinitely large comparatively to its ends.[65] Perhaps, Fresnel realized at the time the connection between the model of light and the concept of coherence, but he did not reveal it.

Besides the conditions of coherence, it was the nature of the destruction of light by interference which continuously attracted Fresnel's attention. In his 1816 memoir Fresnel emphasized that this destruction is only temporary, since

> the rays which have been destroyed by the encounter of opposite vibrations, afterward once again become luminous in the part of the trajectory where the undulations agree, and thus they are able to recover their brightness after having lost it *for a moment*.[66]

The next sentence shows that Fresnel is not implying a transfer of energy from a minimum to a maximum, for

> the undulations when crossing undoubtedly modify themselves at the point of intersection, but afterward their regular movement and their circular form are reinstated.[67]

This is similar to the description of intersection of two waves on a water surface and means that the interfering waves have all their properties restored after they leave the area of intersection. In another obscure remark, however, Fresnel seems to admit that light can *permanently* lose its luminous and other propert-

63 Fresnel, "Mémoire couronné," 286 n.
64 Ibid., 359.
65 Ibid., 285–86.
66 Fresnel, "First Memoir," 27. Italics added.
67 Ibid.

ies.[68] A few months later Fresnel first demonstrated his concern with energy in the explanation of interference. Like many of his contemporaries, he believed that conservation of energy *(vis viva)* is a universal law of nature, and he applied it to optical phenomena. In his explanation of the colors of thin films he said that the coincidence of the maxima seen in the transmitted light with the minima of reflected light is due to the fact that the energy in the incident light is constant, and, consequently, where more light is transmitted less light is reflected, and vice versa.[69] At about the same time, in 1816, Fresnel applied a similar idea to the explanation of the complementarity of colors seen in the ordinary and extraordinary images in a doubly refracting crystalline plate. According to Fresnel, if red light in the ordinary image is completely destroyed by interference, then the image seems to be blue-green. Due to conservation of energy, that part of it which was lost in the ordinary beam will pass to the extraordinary one. The excess of energy in the red part of the spectrum leads to the formation of a maximum, and the extraordinary image is colored red.[70] Fresnel gave his best explanation of the energy redistribution due to interference in a letter to Young in 1819. Taking "Fresnel's mirrors," as an example, he demonstrated that the total energy concentrated in two adjacent bright and dark fringes is constant and the same as if the two light waves did not interfere at all.[71] However, this important result, along with many others related to coherence and energy redistribution, was left unpublished.

VII.3 Third period (1819–1822)

In 1821, Fresnel finally published his theory of chromatic polarization. He gave an analytic expression for the intensity of light due to interference of two light beams polarized in perpendicular directions. He also emphasized the role of conservation of energy in the complementary coloration of two such beams. The most important part of the paper was his explanation of the non-interference of rays obtained from natural light and polarized in perpendicular directions. Fresnel assumed that natural light is a system of waves polarized in all possible directions in such a way that the intensity of light in two arbitrary perpendicular directions is the same. When two such perpendicularly polarized waves interfere, they

[68] Ibid., 27–28.
[69] Fresnel, "Second Supplement," 142.
[70] Fresnel, "Mémoire sur l'influence de la polarisation [August 30, 1816]," *Oeuvres* 1:401.
[71] Fresnel to Young, September 18, 1819, *Oeuvres* 2:748.

produce a mixture of two complementary colors of equal brightness, which is equivalent to white light. The absence of coloration means that the rays do not interfere. When the incident light is polarized, however, the two perpendicularly polarized components have a different intensity and their interference does produce a color.[72]

Fresnel never had enough space available to him in the *Annales de Chimie et Physique* to present his theoretical considerations in detail. Thus when Jean Riffault, the translator of Thomas Thomson's *System of Chemistry*, offered to him to publish in its "Supplement" a popular article on his theory of light, Fresnel readily agreed. This article "De la lumière" appeared in 1822, and it contained Fresnel's best exposition of the principle of interference.

Here Fresnel is more specific and systematic when dealing with subjects that he mentioned only briefly earlier, such as periodicity of light or its discontinuity.[73] The concept of a wave train becomes his major aid in explaining some conditions of coherence, and particularly the limitation of path difference. Imagine two identical wave trains, Fresnel says, moving in the same direction and differing in their paths by a half-wavelength. These wave trains destroy each other everywhere except for a half-wavelength at their ends. Since these ends are very small compared to the length of the wave trains, almost all the wave motion will be destroyed, and a dark fringe will be produced. By adding another half-wavelength to the path difference of the wave trains, Fresnel continues, we obtain a bright fringe and two undestroyed half-wavelengths at the ends; the path difference of three half-wavelengths again produces a dark fringe and three undestroyed half-wavelengths at the ends, and so on. This alternation of dark and bright fringes, Fresnel concludes, is similar to that which follows from the principle of interference as an experimental law.[74] It would have been easy for Fresnel to deduce from this consideration that the fringes' contrast must diminish with an increase of the path difference, since the "undestroyed" light is subtracted from the maxima and added to the minima, but he missed this opportunity. He did not miss, though, another significant result, namely, that when the path difference exceeds the length of the wave train, there cannot be any interference at all.[75]

Another explanation of the limitation of the path difference, according to Fresnel, arises from the compound character of light. He says that even the best monochromatic light available still contains many waves of different frequency, and a small difference in wavelength, when repeated many times, leads to a

[72] Fresnel, "Note sue le calcul des teintes . . .," *Oeuvres* 1:613–15.
[73] Fresnel, "De la lumière," 45, 47.
[74] Ibid., 45–48.
[75] Ibid., 50.

coincidence in the spectra of high orders of maxima of some colors with minima of others. Fresnel believes that lack of perfect monochromaticity is the principal reason for the non-interference of light when the path difference is larger than 50 or 60 wavelengths.[76] The origin of this number is not clear. It seems to be based on Newton's observations of the colors of thin films viewed through a prism; Fresnel cites Newton as having observed thirty rings, although Newton himself estimates between forty and one hundred.[77]

In fact, the two causes for the limitation of the magnitude of the path difference are equivalent, but this became clear only later in the nineteenth century.[78] Fresnel did not see the connection between the upper limit of the path difference obtained from a consideration of the overlapping of the fringes of different colors and the length of the wavetrain, which he evaluated as 1000 wavelengths.[79]

Fresnel's explanation of the necessity of a common light source is similar to that in his 1816 memoir on diffraction, but it is more lucid. According to Fresnel, the particles of luminous bodies experience frequent irregular distortions of their oscillations, which occur non-simultaneously in different particles. Therefore, the phase difference of waves emitted by two neighboring particles changes chaotically and very rapidly, producing in the eye a sensation of a uniform illumination. It is otherwise when both light pencils come from the same center of vibrations, since they experience the changes simultaneously and thus retain a constant phase difference. In this case the effect of interference is permanent and may be perceived.[80]

The role of the size of a light source is one of Fresnel's major points in this article. To simplify his discussion, he distinguishes between the interference of non-parallel rays and that of parallel rays. The first category includes such phenomena as diffraction, the two-slit experiment, "Fresnel's mirrors," and others. He begins his discussion with "Fresnel's mirrors" and derives the expression for the fringes' breadth (7.12).[81] Then he examines the influence of the dimensions of

---

[76] Ibid.
The fact that Fresnel applied both continuous and discontinuous models of light (in other words, wave trains and infinite waves) to account for the same condition of path difference shows that physicists are less rigid in their criteria of "true" solutions than historians. At the time, both models had the same degree of plausibility, and selecting one or the other was a matter of convenience.

[77] Fresnel, "First Supplement," 52. He suggested to use a biprism to observe more than thirty rings. See also Newton, *Opticks*, 222–23.

[78] According to the Fourier-representation of light, a wave train consists of a group of waves of different frequency. The smaller the bandwidth of the spectrum, the longer ist the wave train. Thus, an improvement of the monochromaticity of light is equivalent to enlarging the wave trains. Fresnel's monochromatic wave train can be considered as an approximation of a group of waves. See Françon, *Optical Interferometry*, 8–13.

[79] Fresnel, "De la lumière," 49.

[80] Ibid., 50–51.

[81] Ibid., 54–57.

a luminous body on the formation of the internal fringes. A luminous point creates the brightest (white) fringe where the interfering rays have zero path difference. When the source S is placed symmetrically relative to the plate AB (see Fig. 38) this white fringe is at the center O of the diffraction spectrum. If the

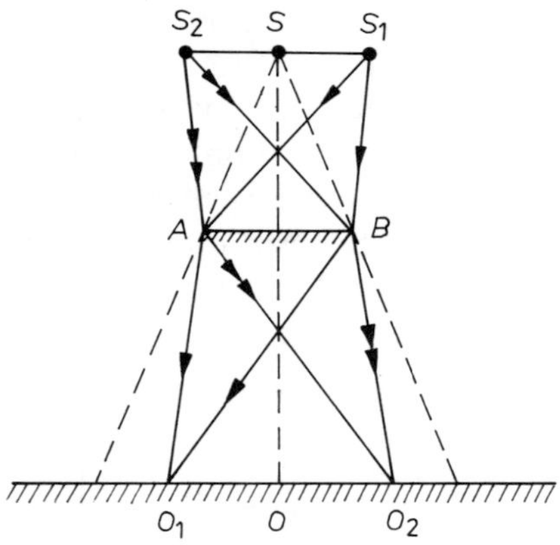

Fig. 38
Fresnel's explanation of the influence of the size of a luminous source on the internal fringes

luminous point moves to S_1, the white fringe moves to O_1 with the rest of the fringes centered around it. If the luminous point moves to S_2, the white fringe shifts to O_2. Thus, when the source of light is an extended body, each point of it froms its own set of internal fringes, and the overlapping of different groups obliterates the fringes. According to Fresnel, the degree of overlapping depends on the angle between the interfering rays:

> That is why in the interference phenomena in which rays cross at noticeable angles, as in all diffraction phenomena, to perceive the effect of the mutual influence of rays it is necessary to employ a very thin luminous body, and so much thinner, as the angle between the rays is larger.[82]

Since, according to eq. (7.12), a larger angle corresponds to narrower fringes, Fresnel's condition may be reformulated as follows: to obtain many narrow fringes, the source must be narrow. This interpretation agrees with Fresnel's claim that the different groups of fringes reinforce each other when their centers

[82] Ibid., 58.

"are very little removed from each other *relative to the fringes' breadth*."[83] In other words, the fringes will be visible as long as the breadth of the body is less than that of the fringes. Fresnel does not demonstrate this condition; it can be shown that this condition is correct if applied to the *angular* breadth of the luminous body and the fringes. The idea is that the radius of the luminous body R is small enough if the shift r between the fringes produced (for instance, in the twoslit experiment) by light coming from its edge and those produced by light from its center is much less than one half of the fringe breadth δ (Fig. 39).[83a]

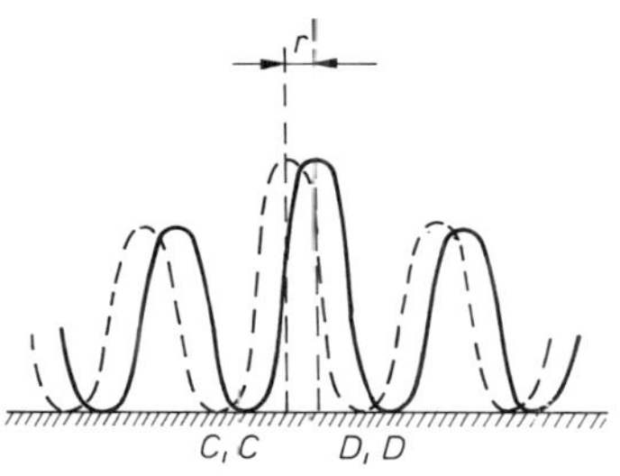

Fig. 39
Consideration of the maximal size of a luminous source

Let us now see how Fresnel deals with the interference of parallel rays. According to eq.(7.12), such rays must produce infinitely broad fringes, and Fresnel seems to claim precisely this:

> When the two systems of interfering waves are parallel, the interval that separates their corresponding points should remain the same over a large part of the wave surface, or, in other words, the fringes must reach an *almost infinite* breadth, and consequently, a significant displacement of the center of undulation does not bring about a noticeable change in the degree of accord or disaccord of their vibrations. That is why, in this case, to perceive the effect of their mutual influence it is no longer necessary to use so small a luminous object.[84]

---

83 Ibid., italics added.

83a Let the interval between the slits A and B (Fig. 38) be $2c$, and the distances from the slits to the luminous body and the screen a and b, respectively. The source of light at S_1, will produce an additional path difference P, as compared to the source at S, such that $P = S_1A - S_1B$. It is easy to show that $P = 2cR/a$, and that $R/a = r/b$. Thus, if $r \ll \delta$, than $R/a \ll \delta/b$, or, the angular diameter of the luminous body must be much smaller than the angular breadth of a fringe.

84 Ibid, 58–59, italics added.

Since the fringes are of "*almost* infinite breadth," Fresnel apparently is considering waves, that are not exactly parallel; or spherical waves of very large radius, which can be considered as parallel in a large region of space. I interpret the "interval that separates their corresponding points" as the path difference. Since each fringe corresponds to all pairs of rays with the same path difference, in this case the fringe will be as broad as the wave surfaces in that part where they are parallel.

When such a wave comes to a transparent plate with two parallel surfaces, the waves reflected from the two surfaces maintain a constant path difference along their surface. For this reason they do not form fringes at the screen T (Fig. 40), for

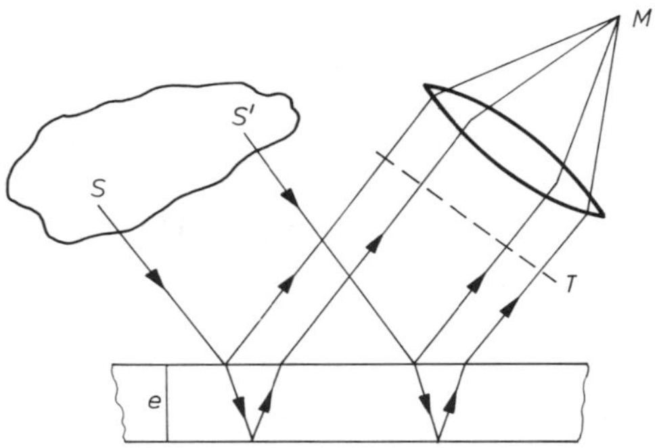

Fig. 40
Interference of parallel rays according to Fresnel

a single broad fringe is equivalent to a uniform illumination. To see fringes, the eye must be accomodated to infinity. Since all rays having the same direction form *one* fringe, the source may be of any size. Fresnel correctly notes that the fringes observed on a thin film are visible because its two surfaces are not perfectly parallel.[85]

In summarizing his discussion of interference of light Fresnel presents for the first time an almost complete list of the conditions of coherence:

F (3) 1822 One must understand now why light rays, although they always exert a certain influence on each other, exhibit it so seldom and in cases so specific; that is, to render it visible, it is necessary: 1) that the interfering rays originate from a common source; 2) that they differ in their paths only by a very limited number of undulations, even when the most sim-

[85] Ibid.

> plified light is used; 3) that they do not intersect each other at too great an angle, because the fringes become so narrow as to be invisible even through the strongest magnifier; and 4) as long as these rays are not parallel and form among themselves a noticeable angle, the luminous object should be of very small dimensions, and so much smaller, as this angle is greater.[86]

Thus, Fresnel's contribution to the concept of coherence consists of: 1) a new condition of coherence for polarized light; 2) an explanation of the role of the source's size; 3) a new explanation of the necessity of a small path difference by means of wave trains; and 4) a generally much clearer explanation of all the conditions of coherence. While Young applied the concept of coherence only implicitly, Fresnel did it explicitly, to the benefit of his readers. Fresnel also was the first to realize the interrelation between the different conditions of coherence: he believed that interference could be observed with different combinations of a source's size, degree of monochromaticity of light, and path difference.

VII.4 Summary

The striking resemblence between Young's and Fresnel's early theories of interference shows that this theory was a natural step in the mathematization of physical optics. Fresnel's theory of interference was superior to Young's in three points. First, it was more sophisticated mathematically in the case of diffraction (1819) and chromatic polarization (1821); second, it agreed better with observation in the two cases; and third, the principal physical concepts of the theory were presented with greater clarity and in more detail (1822). The disadvantages of Fresnel's theory consisted of a smaller number of phenomena explained and lack of exact measurements in cases other than diffraction. Only in 1822 Fresnel explained satisfactorily the meaning of the principle of interference and formulated the conditions of coherence F(3): prior to that date French readers had to rely in this on Arago (A(1), A(2), and A/F) who in turn used slightly modified Young's formulation Y (5).

[86] Ibid., 59. The "completeness" ist relative, of course: the condition of frequency is missing, while having both the condition of direction and that of the path difference is a redundancy. Fresnel realized the advantage of quasi-monochromatic light over white light, and he could do well by introducing the condition of monocromaticity of light, but he missed this opportunity.

Fresnel's early work on interference as represented by his 1816 memoir on diffraction surpassed Young's only in a more detailed mathematics and a greater number of experimental data. As to the precision of measurements, use of hypotheses, and the explanation of the physical aspect of interference, it was no superior to Young's. Fresnel's paper contained a number of errors and obscure passages, and lacked essential details, though some of them were known to him. These flaws are particularly noticeable in his formulations of the principle of interference and the concept of coherence. Of all the conditions of coherence, only that of a common source of light was explained in 1816. The condition of frequency was not even mentioned, and the conditions of direction and of path difference were only applied implicitly. Therefore, it is unlikely that Fresnel's 1816 memoir could have aided those who failed to grasp the principle of interference from Young's works.

Thus, by the standards which historians like to apply to the early nineteenth-century optics, Fresnel's early theory was not superior to Young's. Moreover, none of his contemporaries claimed that it was.

Chapter VIII

Response to Fresnel's principle of interference

According to the accepted view, the wave theory of light was first established in France as the outcome of the struggle between the "undulationists" and the "emissionists." The former group included Fresnel, Arago, Ampère, Dulong, Petit, and Fourier, with Arago playing almost as important a role in the development and promotion of the wave theory as Fresnel himself. Laplace, Poisson, Biot, and other "emissionists" used all means to stop the advance of the newly revived wave theory of light.[1]

This view neglects or misrepresents a number of facts. The debates on the relative advantages of the emission and wave theories of light did take place, but it is not clear whether these debates were in fact the major cause for abandoning the emission theory in favor of the wave theory. It has been suggested that during these debates Fresnel and Arago brought forth new arguments and observations to convince their opponents. The rapid rise of Fresnel's reputation within the Parisian scientific community undoubtedly proves that he did persuade his opponents. The question, however, is of *what* did he convince them?

I will argue that it was the usefulness of the principle of interference rather than the truth of the wave theory. When Fresnel and Arago claimed that some qualitative results of their observations of periodic colors contradicted the emission theory and supported the wave theory, they implied that these results could not be explained without the principle of interference. I will show in this chapter that the principle of interference itself was established by quantitative and not qualitative arguments, that the adoption of this principle became a major preliminary step to the establishment of the wave theory of light, and that this principle was first accepted by a group of influential scientists simply as a means to give a mathematical treatment of diffraction. It will be demonstrated that the only part of Fresnel's theory that was established during his lifetime was the theory of interference, whose acceptance did not necessarily imply the support of the wave theory; and the theory of interference was adopted only for the simplest two-ray model. This allows us to extend a comparison of the acceptance of Young's and Fresnel's theories beyond 1819. It will be shown that the reaction of

[1] Frankel, "Jean-Baptist Biot," 336, 346; "Corpusclar optics and the wave theory of light: the science and politics of a revolutions in physics." *Social Studies of Science* 6 (1976): 155–58, 172; and "J. B. Biot and the mathematization of experimental physics in Napoleonic France," *Historical Studies in the Physical Sciences* 8 (1977): 68. Robert Fox, "The rise and fall of Laplacian physics," ibid, 4 (1974): 112–14. Maurice Crosland, *The Society of Arcueil* (Cambridge, Mass., 1967), 409.

individual scientists to the principle of interference was primarily determined by an interplay between their views on the nature of light and their philosophy of science. In particular, I will demonstrate that the leading French "emissionists" adopted the principle of interference by ignoring its undulatory aspect and emphasizing its role in the mathematization of the theory of periodical colors. It will be also shown that the reaction of the "undulationists" to Fresnel's theory cannot be treated as an *organized* or *purposeful* movement for promoting the wave theory of light.

In general, I will study the response to the principle of interference between 1816 and 1830, but to achieve a proper understanding of Young's and Fresnel's contribution to the concept of coherence, I will examine the comprehension of this concept at a later period too. My discussion basically follows chronological order and begins with the scientist – François Arago – who was the first person in France, after Fresnel, to adopt the principle of interference and the wave theory of light.

VIII.1 Arago

I will show that Arago assisted in the establishment of the principle of interference by his experiments and his personal support of Fresnel, while the role of his theoretical contributions have been greatly exaggerated. Arago first revealed his interest in the nature of light when comparing the velocity of light from various celestial bodies. The emission theory suggested that a difference in the gravity of stars or in their motion relative to the earth could affect the velocity of light, and that rays of different velocity must be refracted differently by a prism. Arago applied this idea in 1806 and in 1810 and found the velocity of light to be constant.[2] To account for the observed uniformity of the velocity of light, Arago could have accepted Young's wave explanation, but he preferred to reconcile the phenomenon with the emission theory by assuming that light from different stars does have slightly different velocities, but within certain limits the eye is unable to perceive this difference.[3]

[2] Arago, "Vitesse de la lumière [read December 10, 1810]" *Oeuvres complètes de François Arago,* 15 vols (Paris, 1854–1859), 7: 548–68; henceforth is cited as *Oeuvres d'Arago* or simply *Oeuvres,* when this is unambiguous. On Arago's 1806 results see *Proces-verbaux* 4: 243–45.

[3] Ibid, 554. He mentioned Young's "Theory of light" only as his source of information for Robison's ideas on the velocity of light from moving sources. In Arago's view, the wave theory could not explain the refraction of light by a moving body. In fact, when the ether is immobile or fully dragged by the moving body, it is a simple mathematical problem.

In 1811 Arago found that a thin film of air or other substance polarized both the reflected and the refracted light in the same plane, whereas a thick plate, according to Malus, polarized them perpendicularly to one another.[4] Arago's paper was to appear in the third volume of the *Mémoires d'Arcueil* early in 1814. However, the occupation of Paris in April 1814 and the subsequent financial problems of the publisher delayed publication until 1817. During this interval a part of Arago's manuscript disappeared from the printer, and this has made it impossible to check Arago's claims about important conclusions relevant to the wave theory and the principle of interference contained in the missing part. It is clear from the remaining text that the lost part included a discussion of different theories of the colors of thin films. In the preface Arago mentioned that among his predecessors was Hooke, whose explanation "has some analogy" with Young's, and he promised to return to Young later in the paper. The published paper does not contain this discussion, however.

Let us examine Arago's claims, which appeared in two subsequent notes summarizing his achievements in the paper on thin films. In the note added to the paper in 1817 Arago says that his experiment gave him the means "to compare the theory of *fits* in the emission and wave systems."[5] He also demonstrated, Arago continues, the insufficiency of the emission theory which can be reconciled with new phenomena only by a continuous addition of new properties of light. He adds that he has no intention to reconstruct the lost piece, because he is short of time, and "the nice experiments of Doctor T. Young and Fresnel's recent observations of the curvilinear propagation of diffraction fringes seem to me to solve the problem."[6] In the second note, written after 1831, Arago states that

> I have established the conformity of my observations with Young's theory of rings, based on the doctrine of interference, particularly as it concerns the angle of polarization.[7]

He also mentions a mathematical demonstration of the equal intensity of fringes produced by a thin plate in reflected and refracted light. This result, according to Arago, is evident in the emission theory but hardly comprehensible in the wave theory. He emphasizes that his paper added new *facts* to those discovered by

---

4 Arago, "Mémoire sur les couleurs des lames minces [read on February 18, 1811]," *Mémoires d'Arcueil* 3 (1817): 323–70 (pp. 323–352 are erroneously paginated as 223–252); references are given with corrected pagination.

5 Ibid., 364.

6 Ibid., 364–65.

7 Arago, "Les anneaux colores," *Oeuvres* 7: 412–13; unpublished.

Hooke and Newton but says nothing about the *theory* of the colors of thin plates, although the missing part contained a special section on this subject.

The two notes seem to contradict one another. Arago implies that he proved (in 1811) the superiority of the wave theory to the emission one and supported the principle of interference, but he does not cite any argument in favor of the wave theory and fails to recognize the importance of the principle of interference for the theory of the colors of thin films. It seems that Arago's retrospective evaluation of his former theoretical views was affected by his later understanding of the problem of the nature of light. In fact, neither the wave theory nor the principle of interference are mentioned in the published paper. Arago's criticism of Newton's theory of fits does not necessarily mean that he supported the wave theory. Many physicists objected to different parts of Newton's optics (especially the theory of fits) but they intended to improve the emission theory rather than abandon it in favor of the wave theory. There is no hint in the paper that Arago adopted the concept of interference, for he believes that the colors of thin plates result from a modification of white light by the first surface of the plate. One of his major results is to show that light polarized perpendicularly to the plane of reflection, when falling on the surface at the angle of complete polarization, produces no fringes either in reflected or refracted light. Since, according to Malus, no light is reflected at this angle, Arago suggests that reflection of light is instrumental for the formation of both types of fringes. However, Arago never mentions that *two* rays are necessary to produce the fringes, and his result could be hardly viewed as supporting Young.

In his papers written in 1811 and in 1812 Arago clearly shows himself a proponent of the emission theory.[8] By 1815, though, he accumulated many objections to it, and he received Fresnel's memoir on diffraction as a means for refuting the emission theory. Apparently, when he met Fresnel in July 1815, Arago expressed a sympathy to the wave theory. However, while informing Fresnel of Young's works, he did not emphasize the role of the principle of interference. Several historians thought that Arago applied the principle of interference to the scintillation of stars as early as 1814,[9] but they were misled by the editors of Arago's

---

[8] Ibid., 336, 360. See also his unpublished papers: "Notes sur les phénomènes de la polarisation de la lumière [1812]," *Oeuvres* 10: 77, 79, 83: "Mémoire sur plusieurs nouveaux phénomènes d'optique [read on December 14, 1812]," ibid., 89; and "Quatrième mémoire sur plusieurs nouveaux phénomènes d'optique [read on December 28, 1812]," ibid., 119.

[9] Frankel, "Jean-Baptist Biot," 296, and "Corpuscular optics," 358–63; also Silliman, "Augustin Fresnel," 163.

posthumous works, for the note on the scintillations was written and published only in 1816.[10]

Besides scientific arguments, Arago had personal reasons to attack such prominent "emissionists" as Biot and Pierre-Simon Laplace (1749–1827). In 1812, following Arago's discovery of chromatic polarization, Biot began to study this phenomenon without Arago's consent. Arago took this to be an invasion of his domain, and his relations with his former friend became very strained.[11] Laplace, in Arago's view, was guilty of establishing a "dictatorship" in the Academy of Sciences, although some contemporaries believed that Arago was not struggling for "democracy" but for power for himself.[12]

In his early works Arago exhibited great experimental skill but little theoretical insight, though he considered himself very competent in theoretical problems. I will show in several examples how these features affected his collaboration with Fresnel. One example concerns the problem of the loci of maxima and minima. Fresnel announced in his "First Memoir" the discovery of a "very remarkable consequence" of his theory, namely, that the external and internal fringes propagate along hyperbolae, and he claimed to have experimentally verified this for the external fringes.[13] Arago became very enthusiastic about this result and suggested that it "may serve to prove the truth of the undulatory theory." [14] He urged Fresnel to improve the precision of his measurements in order to obviate possible objections from the "emissionists." Since Fresnel did not have

[10] J. A. Barral, the editor of the *Oeuvres d'Arago*, presented this note as the "explanation of the phenomenon of the oscillation given to Mr. de Humbodt in 1814 and inserted at the end of the book IV of the *Voyage au régions équinoxicales du nouveau continent*" (*Oeuvres* 7: 97). Barral probably received this information from Humboldt himself, for Humboldt made a similar statement in his "Introduction" to the *Oeuvres*; see 1: XIII. Humboldt's memory was not too reliable in 1854, since only the first volume of his *Voyage* appeared in 1814, whereas the one containing Arago's note was published in 1816. Incidentally, this was not Humboldt's only mistake, for he dated Arago's "glass-screening" experiment to 1818 and 1820 instead of 1816; *Oeuvres* 1: X, XIV. See A. Humboldt, and A. Bonpland, *Voyage aux régions équinoxicales du nouveau continent*, vol. 4, book 4 (Paris: Librarie grecque-latine-allemande, 1816), 285–7. This note could not be published in 1814, for it referred to February 1816 issue of the *Annales*.

[11] Frankel expresses a similar view in "Jean-Baptist Biot," 296. On the Arago-Biot relationship see Crosland, *Society of Arcueil*, 332–35; and Ivor Grattan-Guinness, "Essay Reviews. Recent researches in French mathematical physics in the early 19th century," *Annals of Science* 38 (1981): 669, 676–78.

[12] The members of the Academy of Sciences called Arago "the great elector" for his role in academic elections; see Arago, "Histoire de ma jeunesse," *Oeuvres* 1:99. Fresnel's uncle, Léonor Mérimée, referred to him as "général Arago" for his activity in academic intrigues (*Oeuvres de Fresnel* 2:841). Baden Powell spoke of Arago as the "Napoleon of Science" (*The Edinburgh Review* 104 (1856): 160); and Libri was concerned with his "instinct for domination" (*Revue des deux mondes* 21 (1840): 800. For a different interpretation of the struggle between Laplacians and anti-Laplacians see Fox "The rise and fall," 109–27.

[13] Fresnel, "First Memoir," 19.

[14] Arago to Fresnel, November 8, 1815, *Oeuvres de Fresnel* 1:38.

the appropriate equipment in his village, Arago helped him to obtain a leave of absence to come to Paris where he could work in his laboratory. Arago even verified a number of Fresnel's measurements. Thus, from a referee of Fresnel's paper Arago became his collaborator. He was very much interested in Fresnel's paper as a means to crush the emission theory. Biot and his student Claude Pouillet (1790–1868) were also working on diffraction then, and Arago rushed Fresnel and tried to keep his results secret.[15] On February 26, 1815, Arago hinted at a meeting of the Academy of Sciences that Fresnel's paper "seems to me aimed at making an epoch in science." [16] But only in his report on the paper, read on March 25, 1816, did Arago finally reveal his "secret weapon." He said of the hyperbolical loci of fringes that,

> In the theory of agreements and disagreements it is not necessary to attribute this curvilinear movement *to the light*, . . . we do not know how this peculiar movement coluld be reconciled with the emission hypothesis.[17]

Thus, according to Arago, in the emission theory, the curvilinear loci of fringes require the curvilinear propagation of light. Fresnel, on the contrary, stressed in his "First Memoir" that the hyperbolical propagation of the external fringes does not imply the curvilinear motion of light; and he said nothing about it refuting the emission theory.[18] Moreover, he tried to persuade Arago that this result could not be new, for Young, who had the same theory, could have easily derived the same consequence from it.[19] This was obvious to Fresnel, particularly since Arago himself attributed to Young the same discovery for the internal fringes.[20] Nonetheless, Arago insisted, and Fresnel was in no position to argue with him; consequently, he mentioned in the published paper that the hyperbolical propagation of fringes contradicts the emission theory (without specifying the reasons for this).[21]

In fact, the curvilinear loci of fringes were well known to Newton.[22] Newton believed, though, that each fringe is produced by one ray only, and that rays propagate rectilinearly. To save the phenomena, he assumed that the fringes

15 L. Mérimée to A. Fresnel, December 1, 1815, *Oeuvres des Fresnel* 2: 832–33.
16 Arago, "Note sur la diffraction," 76.
17 Arago, "1816 Report," 86.
18 Fresnel, "First Memoir," 22.
19 Fresnel to Arago, November 12, 1815, *Oeuvres de Fresnel* 1: 61.
20 Fresnel to Young, May 24 1816, *Oeuvres* 2: 738.
21 Fresnel, "Second Memoir," 103.
22 Newton, *Opticks*, Book III, observations 1, 3, 9, pp. 318, 323, 332.

formed at different distances from the body are due to different rays of light.[23] Young informed Arago of Newton's explanation, which made the alleged refutation of the emission theory false.[24] Later Biot made the same point as Young.[25] Nonetheless, Arago continued to assert that the curvilinear propagation of fringes is incompatible with the emission theory of light.[26]

The problem of the loci of fringes was not the only theoretical point misunderstood by Arago. Among other things, he attributed to Hooke the "first elements" of interference.[27] He did not realize the significance of Fresnel's demonstration of the laws of reflection and refraction. This fact in itself can hardly be called unusual, since none of the "undulationists," including Young, appreciated this explanation. However, Arago's reasoning is in some respects remarkable, for he preferred Newton's emission theory of refraction to Fresnel's wave one.[28] He said in the same report that in Fresnel's theory the interference of light is a hypothesis derived from an analogy with beats of sound, while in Young's theory the existance of interference is proved experimentally, and that he, Arago, preferred Young's approach.[29] This phenomenological approach is evident in Arago's description of Fresnel's theory, for he is concerned solely with the principle of interference and ignores the wave hypothesis. Moreover, he focuses only on the mathematical part of this principle, demonstrating how dark and bright fringes are produced by an intersection of circles of a "different sort" and of those of the "same series." This attitude toward interference could not help Arago to comprehend the conditions of coherence. He did not understand the influence of the dimensions of a luminous body and believed that Young's two-slit experiment could have been performed in Grimaldi's manner.[30] Arago never applied the Huygens–Fresnel principle, and for many years he refused to accept Fresnel's idea of transverse light waves.[31] Thus his wave theory was, in fact, reduced to a simple

---

[23] Ibid., 332, see also Fig. 1 on p. 321.

[24] Young to Arago, January 12, 1817, *Misc. Works* 1: 381–82.

[25] Biot, "Additions à l'optique," in Fischer, *Physique mécanique*, transl. from the German by J. Biot, 3rd ed. (Paris, 1819), 418.

[26] Arago, "Thomas Young," *Ouevres* 1: 292.

[27] Arago, "1816 Report," 82. Hooke's discovery may be considered to be the selection of interfering rays for thin film phenomena; it is useful, however, only to a person who already knows the principle of interference.

[28] Arago, "1816 Report," 82–83. Just a few months earlier, Arago severely criticized Newton's theory of refraction; Arago and Petit, "Mémoire sur les puissancees réfractives et dispersives de certain liquides et des vapeurs qu'ils forment," *Ouevres d'Arago* 10: 123–31.

[29] Arago, "1816 Report," 85.

[30] Arago, "Influence mutuelle de faisceaux lumineux," 124.

[31] For instance, he did not mention transverse waves in Fresnel's biography in 1830; see *Oeuvres d'Arago* 1: 107–85.

theory of interference of Young's type. Arago's ultimate goal had been a refutation of the emission theory, rather than building a wave theory, and he wanted to achieve this by multiplying the phenomena which could be accounted for by the principle of interference. Therefore, Arago's contribution to the establishment of the wave theory was indirect and exclusively experimental.

Most of Arago's observations of interference, such as "Fresnel's mirrors," the interference of polarized light, and the "glass-screening" experiment, were conducted together with Fresnel. From the "glass-screening" experiment Arago derived a method of measuring small differences of refraction in two media which he decided to use to discover whether humidity of the air influences atmospheric refraction. He invited Fresnel to join him, and together they built the first interferometer.[32] The only phenomenon of interference which Arago studied independently was the scintillation of stars. He suggested that different rays of light from a star reached the eye by slightly different routes after passing through layers of the atmosphere of different density. The motion of the atmosphere continuously changes the density along the routes of the interfering rays, and this leads to a variation of the path difference and intensity of perceived light. This explanation was qualitative, and there is no evidence that Arago himself ever developed any quantitative theory of this or any other interference phenomenon.

Undoubtedly, Fresnel benefited from Arago's patronage in practical matters. Arago helped him to settle in Paris, provided him with a laboratory, and published his papers. Without this assistance Fresnel perhaps would have never become a scientist. Unfortunately, Arago tended to extend his patronage too far by trying to guide Fresnel in his research. As we have seen, Arago was not suitable for such a role: he promoted only that part of Fresnel's theory that he himself considered worthwile, that is, the principle of interference.

It appears that theoretical differences between Arago and Fresnel played a role in the delay of the reports on Fresnel. Although Arago had the opportunity to publish all Fresnel's papers in his journal, between 1816 and 1819 he published only three out of total of ten. In the published papers Fresnel was usually limited in the space available to him and could not discuss his theoretical views

[32] Arago, *Oeuvres* 10: 315–20.

in detail.[33] The accounts of the joint investigations of Fresnel and Arago appeared under Arago's name, and Fresnel did not receive full credit for his contribution. For instance, only a passing remark at the end of the note on the "glass-screening" experiment relates that Fresnel predicted the result of the observation with a very thin plate.Arago creates the impression that the theory was applied (it is not said by whom) *after* he completed the whole set of modifications of this experiment, while in fact the results of all these observations were predicted by Fresnel's theory.[34] When reading this paper to the Academy of Sciences Arago totally omitted the passage on Fresnel's prediction.[35] We do not know exactly what Arago wrote to Young about the study of the interference of polarized light (this letter is not preserved), but in his reply Young called Arago the discoverer of the non-interference of perpendicularly polarized rays (in fact, Fresnel had attained this result before Arago joined him in this investigation),[36] and

[33] That Fresnel lacked space was acknowledged in an anonymous note "Question de physique, proposee par l'Acad. Roy. des Sciences de Petersbourg," *Bulletin des Sciences Mathématiques* 7 (1827): 212. I attribute this article to Arago. If Arago had better understood the importance of Fresnel's wave theory and was interested in its promotion, he would have given more space to him in the *Annales*, for Arago knew that Fresnel could not afford to publish a book.
It is difficult to establish all the reasons for not presenting reports on some of Fresnel's memoirs and delaying them on others. This was not unusual for the Academy: the academicians were overloaded with the reviewing of incoming papers and inventions and tried to skip some of the reports. Nonetheless, when the reviewers considered the paper important, they completed their reports rapidly. For instance, it took only a month to prepare reports on Berard's 1812 paper on infrared rays (Biot, Berthollet, and Chaptal), on Pouillet's 1815 memoir on the colors of thick plates (Poisson and Ampere), and on Despretz's 1817 paper on heat (Fourier, Gay-Lussac, and Thenard; see *Procès-verbaux* 5: 130, 146, 594, 6: 8, 234, 238). Between 1815 and 1823 Fresnel submitted to the Academy 16 papers on optics, and only 3 reports on them appeared, two of which were delayed by four months, and one by three and a half years after the last supplement to the paper was submitted. According to Biot, in some cases Fresnel did not want his papers to be reviewed; see "Note de M. Biot sur des mémoires des Fresnel qu'on croyait egares," *Comptes Rendus* 22 (1846): 406–7). When two lost papers of Fresnel ("Mémoire sur la réflexion de la lumière [read on November 15, 1819]" and "Mèmoire sur les couleurs developpées dans les fluides homogènes par la lumière polarisée [read on March 30, 1818]") were found, Biot and Arago discovered, to their surprise, that they had been nominated to examine these papers. Biot hinted that Fresnel probably withdrew the papers immediately after their formal registration, to preserve a new field of research for himself. However, the first paper appeared in December 1819 in the *Annales*, while some results from the second memoir, together with those from Fresnel's 1816 paper on chromatic polarization, were published by Biot himself in a note to his "Mémoire sur les rotations que certaines substances impriment aux axes de polarisation des rayons lumineux [read on September 22, 1818]," *Mém. Acad.* 2 (1817): 133–36 (see *Oeuvres de Fresnel* 1: 495–99, 533–37, 589 n). Perhaps, Fresnel simply gave up hope that his papers would be reported in due time, and he tried other routes to inform scientists of his discoveries. I do not see any personal motives in Arago's policy of reporting on Fresnel and publishing his papers. He did his best for the advances of science, as he understood it. Unfortunately for science, Arago's grasp of the role of mathematical theories in contemporary optics was not adequate.

[34] Arago, "Note sur la diffraction," 76–77.

[35] Arago, *Oeuvres* 10:313–15.

[36] Arago to Young, January 12, 1817, 38.

he also announced Arago's authorship in his "Chromatics" in 1817.[37] It is amazing that French readers learned about this French discovery two years later than the British.

Whatever Arago's reasons for doing such injustice to his friend might have been, such acts evidently delayed the establishment of the principle of interference and the wave theory. Arago later claimed that he did not emphasize Fresnel's undulatory views in his reports on Fresnel's papers in order to prevent a controversy on the nature of light that could have negatively affected his scientific career.[38] However, his own optical papers of that time likewise had nothing on the wave theory. Arago's position can be better understood from the debate following his 1816 report.

VIII.2 Reception of Fresnel's first paper

In the conclusion of this report read on March 25, 1816, Arago suggested that the Academy not only expressed its satisfaction with Fresnel's work and recommended it for publication in the *Recueil des Savants étrangeres*, as was the routine, but also

> without stating anything about the merit of the hypothesis which he [Fresnel] has examined with such sagacity, it [the Academy] can engage this able physicist to apply it, if possible, to other phenomena.[39]

The minutes of this meeting of the Academy provide interesting details of the discussion of Arago's report.[40] Although all academicians, except for Arago, were then the "emissionists," Fresnel's adherence to the wave theory did not excite many objections. Only Laplace regretted that Fresnel abandoned the emission theory which, he felt, had explained so many optical phenomena. However, Arago's proposal to encourage Fresnel to apply his "hypothesis" provoked an animated debate.

Traditionally, the Academy approved only the *conclusion* of each report, which was to be free of judgements on the author's hypotheses, since the

[37] Young, "Chromatics," 287.

[38] Arago, "Examen des remarques de M. Biot," *Oeuvres de Fresnel* 1: 593; see also Verdet, "Introduction," LXXXVI; and Whewell, *History* 2: 101, 114–16.

[39] Arago, "1816 Report," 87.

[40] "Mèlanges. Notice des Séances de l'Acad. R. des Sci. de Paris," *Bibliothèque Universelle* 1 (1816): 319–21.

Academy did not support one hypothesis over any other. Arago violated this unwritten rule and was reminded of this. Jacques-Alexandre Charles (1746–1823), Professor of Physics at the *Conservatoire des Arts et Métiers,* and Biot said that the Academy never pronounced on hypotheses and recommended retaining this custom. Laplace urged the derivation of laws from phenomena instead of the invention of new hypotheses. He emphasized that a law does not depend on an hypothesis and recalled Huygens' theory of double refraction, which was originally based on the wave theory but afterwards was reformulated to fit the emission theory. He demanded that the conclusion of Arago's report be changed to urge Fresnel to discover new laws rather than to extend his hypothesis. Arago replied that he was misunderstood, since his term "hypothesis" referred to the principle of interference, rather than to the wave theory, and that this principle was nothing but a "geometric construction presenting the facts in a very exact way, which could be reconciled with the wave theory." This clarification was received approvingly. Poisson commented that he considered Fresnel's memoir to be a "statement of fact, a construction of phenomena," which does not claim to explain *how* the luminous oscillations are produced.[41] André-Marie Ampère (1775–1836) said that,

> It was by a single application of his system of gravity to the moon that Newton verified Kepler's three laws. It was afterwards recognized that the same cause explained the parabolic movement of comets, the floods and the ebbs in a sea, etc, and consequently it became more and more probable. Mr. Fresnel's *construction* of the intersecting circles is still derived only from one phenomenon, but if he were to succeed in extending it to others, it would bring them all under one *law*; and although I have always subscribed to the emission system, the conclusions of the Report seem fine to me.[42]

Adrien-Marie Legendre (1752–1833) recalled how in 1740 the Academy admitted both Cartesians and Newtonians and urged the report be accepted in its original form. Finally, the Academy approved Arago's report without any changes.

Three interesting points emerge from this discussion. First, Laplace's power in the Academy was very limited, for the failed to win an argument in which he had tradition on his side.[43] Secondly, the leading French physicists and mathematicians were more concerned with mathematical laws derived from observation

[41] Ibid., 320.
[42] Ibid., 320–21. Italics added.
[43] Although Arago clarified his term "hypothesis," the readers of his *Annales*, where the report appeared, never learned about this.

than with hypotheses. Finally, they approved Fresnel's principle of interference as such a mathematical law, despite its association with the wave theory. This response will not appear unusual, however, if we take into account the methodological views of these men.

Laplace, the leader of the French school of mathematical physics, formulated his scientific method in the following way:

> To ascend from phenomena to the laws and from laws to the forces is . . . the true course of natural philosophy.[44]

This was Newton's program, and it was quite popular among the best French scientists as early as 1807, when it was proclaimed to be a program of the Society of Arcueil.[45] When historians discuss Newton's program, they usually concentrate on its ultimate goal, that is, the reduction of all phenomena in nature to central forces. However important this approach to the early nineteenth-century science was, the first part of Newton's program ("from phenomena to the laws") was of no less significance. First, it was a call for the mathematization of physics. Secondly, in some branches of physics, such as optics, the discovery of the laws of phenomena was not an intermediate step, but in fact the final product of research. Biot, for instance, said that the hypothesis of short-range forces in optics could not be suggested by phenomena, since these forces cannot be discovered experimentally; it was invented by reasoning, and its consequences were calculated and confirmed by observations.[46] Physicists realized that in optics they must be content with the discovery of mathematical laws, while leaving the completion of Newton's program for the future.

Laplace considered hypotheses to be only a means to establish mathematical laws of nature.[47] He believed that hypotheses and laws are independent of one another, and that the same law may be derived from different hypotheses, as he demonstrated with Huygens' theory of double refraction. Whether influenced directly by Newton, or indirectly through Laplace, the leading French mathema-

---

44 "Laplace to T. Young, October 6, 1817," *Misc. Works* 1:374.

45 *Mémoires d'Arcueil* 1 (1807): III. Laplace, Biot, Gay-Lussac, Arago, Malus, Berard, Poisson, and Dulong were members of the Society of Arcueil.

46 Biot, "Sur l'aimantation imprimés aux metaux par l'électricité en mouvement," *Journal des Savans* 1821:234.

47 He applied his hypothesis of short-range forces to atmospheric refraction, double refraction, and capillarity; see his *Traité de mécanique céleste*, vol. 4, Supplement (Paris, 1805), 1–2. He also used more specific hypotheses, such as the assumption that the velocity of the extraordinary ray in Iceland crystal depends exclusively on the angle between the ray and the optical axis of the crystal; see "Sur le mouvement de la lumiere [1809]".

ticians and physicists shared these views. Poisson, for instance, when discussing propagation of light and heat utilized both the wave and emission theories, for he was guided only by convenience of a mathematical solution of these problems.[48] In the early nineteenth century, Biot was the leading French champion of mathematical science against speculative "systems" which were then still quite popular. According to him, the goal of the physical sciences is a search for mathematical laws of phenomena, rather than for primary causes.[49] He allowed the use of imponderable fluids, but solely as a convenient hypothesis, which may be modified or abandoned if the facts contradict it.[50] Biot distinguished the emission or wave *system* from a *theory*. By "theory" he understood a set of mathematical relations between exactly measured observations.[51] Biot realized his ideal of a physical theory in his theory of "mobile polarization." Though he utilized "light molecules whose axes oscillate between two limits," he never identified this model with reality.[52] Following Newton, he refused to discuss the cause of fits of easy reflection and refraction and took them as a "fact." Biot's theory was built on two concepts: the periodicity of light and the assymetry (or transversality) of light about the direction of its propagation. His attempts to mathematize these concepts were important, though they appeared rather awkward in the emission theory of light.

These examples prove, contrary to the belief of some historians, that a positivistic attitude in physics was not born under the pressure of new anti-Newtonian ideas, such as the wave theory of light. Nor were these views limited to the "emissionists." Ampère and Jean-Baptiste-Joseph Fourier (1768–1830), who are considered to be the major figures of the anti-Newtonian movement in the 1820's, held to the same views, which they had adopted long before the wave

[48] As early as 1819, when the wave theory of light was unpopular, Poisson applied it in his theory of refraction; see his "Mémoire sur les mouvements des fluides élastiques," *Mem. Acad.* 2 (1817): 380–85. But in 1835 when the wave theory of light was almost generally adopted, and the idea of the vibratory nature of heat received a support, Poisson based his theory of heat on the emission hypothesis; see his *Théorie mathématique de la chaleur* (Paris, 1835).

[49] Biot, "Sur l'esprit de système [1809]," in his Mélanges scientifiques et littéraires, 3 vols (Paris, 1858), 2: 112–13.

[50] Ibid, 114.

[51] Biot, "Lettres à Sophie sur la physique [1811]," ibid, 254.

[52] Biot, "Mémoire sur un nouveau genre d'oscillation," 68–71, 138.

theory of light was widely accepted.[53] Thus, it should not be surprising that in responding to Fresnel's paper the academicians ignored the problem of the nature of light or considered it as subordinate to the possibility of a mathematical description of optical phenomena. This explains why there was almost no objections to Fresnel's paper: his mathematical approach was commended, but the topic was too specific, and only Biot was sufficiently familiar with it to comment on physical points. His remarks deserve a separate discussion.

Biot and Pouillet had studied the diffraction of light since the summer of 1815. Biot presentend their preliminary results to the Academy of Sciences on October 9, 1815, two weeks before Fresnel. The work was completed by early March 1816, and on March 4 Biot began reading his paper, which he finished on March 25.[54] Biot and Pouillet studied diffraction on a parallel slit and a circular aperture. They established a quantitative relation between the positions of maxima and minima of different orders. They extended to diffraction Newton's law that a fringe's width is proportional to the interval of fits in the given medium. One of their results was the same as Fresnel's, namely that the fringes' positions do not depend on the substance of the diffractor. However, contrary to Fresnel, they concluded that the loci of maxima and minima were straight lines.[55]

During his study of diffraction Biot heard rumors about a "remarkable work" on the same subject presented to the Academy.[56] Nonetheless, due to the secrecy

---

[53] In 1827 Ampère wrote: "First, to observe the phenomena, varying circumstances as much as possible, and to supplement this initial work with precise measurements, to derive from them general laws based solely on observation; then, independently of all hypotheses on the nature of the forces which produce the phenomena, to derive from these laws, the mathematical value of these forces, that is, the formulae representing them, – that was the course that Newton followed. Generally, in France this course has been adopted *by the scientists to whom physics owes its recent immense progress*, and it was this that guided me in all my researches on the phenomena of electrodynamics" ("Mémoire sur la theorie mathématique des phénomènes électrodynamique," *Mém. Acad.* 6 (1827): 176, italics added). In his theory of heat developed between 1807 and 1821 Fourier proceeded from similar premises: "Primary causes are unknown to us; but are subject to simple and constant laws, which may be discovered by observation, the study of them being the object of natural philosophy. Heat, like gravity, penetrates every substance of the universe, its rays occupy all parts of space. The object of our work is to set forth the *mathematical laws* which this element obeys" ("Analytical Theory of Heat," trans. by A. Freeman, (New York, 1955), 1, italics added); and also: "Of the nature of heat uncertain hypotheses only could be formed, but the knowledge of the *mathematical laws* to which its effects are subject is *independent of all hypotheses*; it requires only an attentive examination of the chief facts which common observations have indicated, and which have been confirmed by exact experiments" (ibid, 26, italics added).

[54] An abstract of Biot and Pouillet's paper was published as "Recherches sur la diffraction de la lumière," *Bulletin de Société Philomatique* 1816: 60–61. The complete memoir appeared as "De la diffraction de la lumière," in the "Supplément à l'Optique" to Biot's *Traité de physique* 4: 743–75. It will be cited as "Pouillet-Biot memoir."

[55] Ibid., 761.

[56] *Bibliotheque Universelle* 1 (1816): 307.

imposed on Fresnel's work by Arago, Biot could hardly have learned anything about it before February 26, 1816, when Arago read the paper on the "glass-screening" experiment and referred to Young's 1803 paper. A week later, Biot demonstrated his familiarity with the principle of interference while reading his paper before the Academy:

> Young seeks to explain the phenomenon [of diffraction] by the wave theory, while the author prefers the method of Newton, who after having well observed and studied the phenomenon came upon his *fits of easy and difficult reflection* (as simple facts); results, to which one will always revert whatever system one adopts. In Young's wave system these oscillations can mutually influence each other, in such a way as to double or destroy their forces, according to whether their directions coincide or oppose one another.[57]

Thus, Biot states that diffraction cannot be explained without the concept of periodicity of light, and that in the emission system this concept coincides with that of "fits." In other words, Biot is attempting to conceive a theory of diffraction which is independent of a hypothesis of the nature of light. It seems that he considers Young's and his own explanations of diffraction as equal alternatives.

During the discussion of Biot and Pouillet's memoir, Arago claimed that Young did not assume a vibrating medium but only compared interference of light with beats of sound.[58] This misinterpretation of Young illustrates Arago's desire to maintain the phenomenological status of the principle of interference. Biot replied, just as Gough had done earlier, that beats of sound could be explained without interference by assuming that the third sound is only a *sensation*, which appears in the ear and does not have any specific source of vibrations outside. No objections to Biot followed, for the academicians apparently were not prepared for such a debate. How little they understood the mechanism of interference can be seen from Poisson's comment. He recalled an incident in which after a simultaneous firing of many guns, no sound was heard for some moments. Apparently he thought this disappearance of sound was similar to that which periodically occurs in the beats of sound.[59] During the discussion of Arago's report on Fresnel, Biot remarked that since the third sound can be accounted for without any concept of interference, the analogy between the beats of sound and interference of light cannot demonstrate the wave nature of light.

---

[57] Ibid., 305.

[58] *Bibliothèque Universelle*, 305–306.

[59] Ibid., 306.

He also added that Fresnel's observations, performed with white light, could not be sufficiently exact and remarked that, "in many cases, after inflection rays propagate rectilinearly."[60] This is obviously an objection to Fresnel's result, but it is not clear whether Biot meant rectilinear loci of fringes or rectilinear propagation of light. He presented more detailed objections in the published version of his paper on diffraction, which appeared *before* Fresnel's.[61]

In explaining the production of colors by a parallel slit Biot follows Newton and assumes that after diffraction light rays propagate rectilinearly, and that fringes formed at different distances from a body are due to different rays. He adds, though, two innovations. First, he assumes that when light passes near a body, it experiences at various distances from the body periodical condensations and rarefactions, which produce a system of straight light beams separated by dark intervals. Since each edge forms such a system of beams, a parallel slit produces two intersecting systems, and Biot supposes that bright fringes appear at the intersection of two light beams. In this way he is able to illustrate the periodicity of fringes and the variation of the number of fringes visible at different distances from the body (see Biot's diagram in Fig. 41). One must remember though that the straight lines in this figure are not the loci of maxima and minima as Fresnel understood them. Biot does not connect the positions of the fringes of the *same order* (counting from the edge of a body), as Fresnel did. Therefore, the comparison of the two cases is meaningless, and Biot's alleged objection to Fresnel's law of the loci of maxima and minima (if this is what he meant) is invalid.[62]

Although Biot, like Young and Fresnel, considers the fringes to be produced by an intersection of *two* rays, his idea is totally different. His rays only amplify and never destroy one another, so that Biot's system of fringes is similar to Newton's, where each fringe is formed by one ray. Moreover, Biot states quite explicitly that rays *cannot* influence each other in Young's and Fresnel's way. He presents two arguments in favor of this view. First, he asserts that an intersection of monochromatic rays does not change their color; and since white light is a mixture of different colors, each of which is unaffected by the crossing, the intersection of two white beams cannot produce any coloration.[63] It is not surprising that Biot did not understand the formation of colors by interference of white light,

[60] Ibid., 319.

[61] The Academy received Biot's *Traité de physique* on April 22, 1816 and the *Annales* with Fresnel's paper on May 13, 1816. Apparently, Biot had an opportunity to make a few additions in his article after hearing Arago's report on March 25. In his book he mentioned that Fresnel's memoir was still unpublished. Without naming Fresnel, Biot criticizes the principle of interference and the curvilinear propagation of fringes (he is not very clear on the latter point); Pouillet-Biot memoir, 761–3, 774.

[62] Pouillet-Biot memoir, 761.

[63] Ibid., 763.

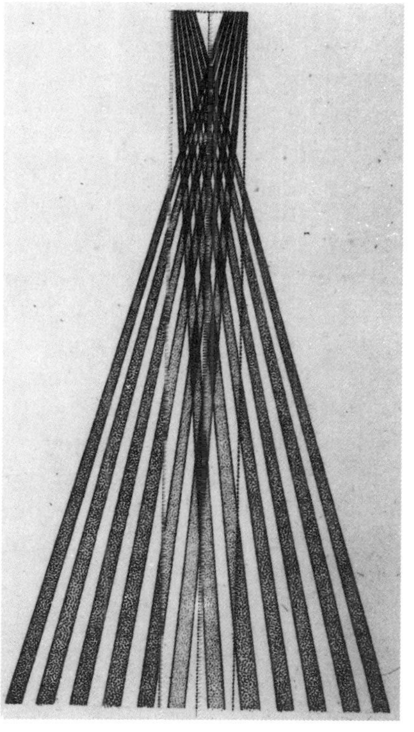

Fig. 41
Loci of maxima and minima in diffracted light, according to Biot (from his *Traité de physique* v.4, Fig. 83)

since Arago had said nothing about it. Second, Biot considers diffraction by two bodies whose shadows intersect on a screen and argues that since the external fringes of these bodies retain their shape and position after crossing, the intersecting rays do not disturb one another.[64] He believes that interference is not the only way to explain the production of the internal fringes and suggests to mathematize Dutour's model of diffraction in the ether atmosphere.[65]

The first comments on Fresnel's 1816 memoir on diffraction were based on Arago's account of it. Let us now turn to the response to the published memoir. A few days after the appearance of Fresnel's paper, Ampère wrote to his friend that

[64] Ibid., 771, 774.
[65] Ibid., 775.

this paper contains the "true theory of light." [66] It is not clear how far Ampère's conversion to the wave theory went at that time. The only thing known is that in 1816 he called Fresnel's attention to transverse waves.[67] It seems that Ampère developed his own interest in the wave theory no earlier than 1824, and only in 1828 did he make his contribution to it.[68]

Several comments on Fresnel's paper appeared in 1817. In France, the only response was from Biot, who mentioned Fresnel's work on diffraction in his physics textbook.[69] After reading Fresnel's paper, Biot stopped questioning the quality of his observations and emphasized their "extreme precision." He remarked that Fresnel's results agree remarkably with the ideas presented long ago by Young, which were derived from an "undulatory movement."

The most important British review came from Young, but it was not very encouraging for Fresnel. Young claimed that one cannot find in Fresnel's memoir "a single *new* fact . . . of the least importance." [70] Young was certain that his principle of interference did not need any additional support from Fresnel, whose work was just a repetition of his own. He did not recognize the importance of Fresnel's discussion of reflection and refraction of light. Nor did he give him credit for the experiment now known as "Fresnel's mirrors," because he probably considered it to be a version of his own two-slit experiment. Thomson mentioned Fresnel's memoir in his review of the physical sciences for 1816 as an "important paper." [71] It is not clear whether Brewster's brief summary of his old observations on diffraction was connected with Fresnel's paper.[72] He stated that the deviation of rays passing near a body does not depend on the body's substance, so that diffraction is an inherent property of light. This communication appears to be a priority claim and might have been a response either to Fresnel or Biot.[73]

In Germany, Fresnel's work attracted the attention of Johann Schweigger (1779–1857), the editor of the *Journal für Chemie und Physik*, who defended the

---

[66] Ampère to Ballanche, May 19, 1816, *Correspondance du grand Ampère*, 3 vols. (Paris, 1936), 2: 511.

[67] Fresnel, *Oeuvres* 1:394 n.

[68] Ampère to Faraday, early in 1825, *Correspondance du grand Ampère* 2: 675; "Mémoire sur la détermination de la surface des ondes lumineus dans un milieu dont l'élasticité est différent suivant les trois directions principales . . .," *Annales de Chimie* 39 (1828): 113–145

[69] Biot, *Précis élémentaire des physique*, 2 vols. Paris, 1817), 2: 496.

[70] T. Young to M. F. Arago, January 12, 1817, 743.

[71] Thomson, "Improvement in physical science during the year 1816. Optics," *Annals of Philosophy* 9 (1817): 3–4. Arago informed Fresnel that his paper was praised by John Playfair (A. Fresnel to L. Fresnel, Juli 19, 1816, *Oeuvres* 2: 835). Neither the source, nor the exakt content of this comment is known, but Arago probably met Playfair in Paris some time before July 19, 1816.

[72] "Proceedings of the Royal Soc. of Edinburgh, June 17, 1817," *Quarterly Journal of Science* 2 (1817): 207.

[73] Brewster corresponded with Biot and Young. Young received several copies of Fresnel's memoir from Fresnel and Arago, and possibly he sent one of them to Brewster.

independence of Fraunhofer's discovery of the dark lines in the solar spectrum of Wollaston's and Young's earlier observations.[74] He disagreed with Arago's suggestion that these lines resulted from an absorption of light.[75] In Schweigger's view, Fraunhofer's dark fringes were of the same kind as those observed in experiments on interference and diffraction by Young, Arago, and Fresnel.

Between 1816 and 1819 Fresnel's scientific reputation was steadily growing in Paris, and it seems that his work on interference was the major reason for this. After his 1816 memoir on diffraction, he published only two short papers which did not concern the principle of interference.[76] However, he was allowed to read to the Academy of Sciences two memoirs on the interference of polarized light.[77] In May 1817, Arago suggested to Fresnel to apply for membership in the Academy. It is interesting to note that Fresnel received only one vote, although two of his friends and allies – Arago and Ampère – were present.[78] In the spring of 1818 Ampère and Magendie offered to put his name on the election list in the Philomatic Society. A vacancy occured in April 1819, and Fresnel was elected.[79]

Young's prestige in France grew as fast as Fresnel's. On February 20, 1816, Biot wrote that,

> his [Young's] testimony has no more weight here than the authority of Aristotle had against Galileo's observations on gravity.[80]

This attitude changed rapidly. A few months later, Arago and Gay-Lussac visited Young in London, and Arago became his admirer and champion. In the spring of 1817, Young was warmly received in Paris. In January 1818, Young became a corresponding member of the Academy, winning the election over Brewster, Leslie, and Venturi.[81] Apparently the rise of Fresnel's and Young's status in France

---

[74] Schweigger, "Auszug aus den Verhandlungen in der Mathematisch-Physikalischen Clase der kon. Akad. der Wiss. zu München," *Journal für Chemie und Physik* 19 (1817): 78–81.

[75] Arago, *Annales de Chimie* 4 (January 1817) 97 n.

[76] One of them was "Extrait d'une lettre de Fresnel à Arago, sur l'influence de la chaleur dans les couleurs developpées par la polarisation" (*Oeuvres* 2: 637–39). Fresnel made a few qualitative observations without a knowledge of Brewster's similar experiments. Another was "Lettre à François Arago, sur l'influence du mouvement terrestre dans quelques phénomènes d'optique" (ibid., 627–36). This was a very important work, but its significance was not realized until the 1850's.

[77] On November 24, 1817 Fresnel read his "Mémoire sur les modifications que la réflexion imprimé à la lumier polarisée," (see *Procès-verbaux* 6: 237), which was reported by Arago in June 1821 and published at the same time. On March 30, 1818 he read the "Mémoire sur les couleurs developpées dans les fluides homogènes par la lumiere polarisée," (ibid. 305), which apparently only Biot was familiar with (see n. 33, above).

[78] *Procès-verbaux* 6: 186–87.

[79] A. Fresnel to L. Fresnel, April 23, 1818: *Oeuvres* 2: 846.

[80] Biot to Brewster, February 20, 1816: Bibliotheque de l'Institut, Paris, Ms. 4895, no. 70. See n. 101, Ch. VI.

[81] *Procès-verbaux* 6: 253, 260.

was to a considerable degree due to their work on interference of light. In 1817, the principle of interference became quite popular among Parisian academicians, as is evident from Young's remark in his article "Chromatics" that:

> many of the most strenuous advocates for the projectile theory have been disposed, especially since the experiments of Mr. Arago and Mr. Fresnel, to admit the truth of all the calculations, in which this law of interference has been employed.[82]

This was written late in 1817 or early in 1818, but Young had obviously formed this opinion when he was in Paris in the spring of 1817. At about the same time the Paris Academy of Sciences officially expressed its recognition of the concept of interference.

VIII.3 Contest on diffraction

On March 17, 1817 the Academy of Sciences announced its next biannual contest in physics. The program of the contest required:

> 1. To determine by precise experiments all effects of diffraction of direct and reflected light rays when they pass separately or simultaneously near the extremities of one or several bodies of limited or indefinite extent, taking into account the intervals between these bodies, as well as the distance from the luminous body from which the rays emanate.
> 2. To derive from these experiments by mathematical induction the motions of the rays in their passage near bodies.[83]

The commission appointed to examine the submitted memoirs consisted of Laplace, Poisson, Biot, Arago, and Gay-Lussac. The story of the "battle" between the "undulationists" and the "emissionists" in the commision has been a favorite topic for historians. Nonetheless, it has not yet been explained how Arago was able to win over Laplace, Poisson, and Biot and secure the prize for Fresnel. I will bring forth evidence to show that there was in fact no reason for the commissioners to differ about the outcome of the contest.

---

[82] Young, "Chromatics," 279–80.

[83] *Procès-verbaux* 6: 165. Due date was August 1, 1818.

Indeed, the program of the contest required a mathematical theory of diffraction verified by observation. Although the program was formulated in terms of the emission theory (Biot was its principal author), there were no limitations set upon the hypotheses adopted. It is important to note a passage from the preamble of the program stating that the internal fringes are produced "when rays *simultaneously pass by two sides* of a very narrow body."[84] Such an explanation was unheard of before Young, and the emphasis on the necessity of *two* rays to produce fringes is an implicit recognition of the principle of interference.

According to Arago's report, the commission was attracted to Fresnel's memoir by "the exactitude of observations, as well as the novelty of results."[85] The "novelty" is obviously not related to the wave theory, since this theory is not mentioned in the report. Even the Huygens-Fresnel principle is described without any reference to waves, just as a mathematical construction which provides an analytical expression for the intensity and location of fringes. The report emphasizes the precision of Fresnel's measurements and the agreement between his theory and observations with respect to both fringes' positions and intensity. Thus, the report shows that Fresnel achieved exactly what was required: he derived a mathematical law of diffraction and confirmed it experimentally. The theory was very general, since Fresnel's integral apparently could be applied to any case of diffraction, and the precision of his measurements surpassed the earlier work in this branch of optics. Therefore, the commision had to be satisfied with the quality of Fresnel's work. Let us now examine whether a difference of views on the nature of light could have affected the commissioners' attitude toward Fresnel.

After a debate on the Huygens-Fresnel principle, late in 1818, Fresnel said of Laplace:

> although he disapproved of my method of consideration in several points, it hardly affected his interest in my research.[86]

In 1822, Laplace announced at a meeting of the Academy of Sciences that Fresnel's researches on double refraction (based on the concept of transverse light waves) "surpass all others communicated to the Academy for a long time."[87] Biot was very impressed with Fresnel's memoir, and he intended to include the Huygens-Fresnel principle and its applications in his course of physics at the *Collège*

---

84 Ibid., italics added.

85 "Rapport fait par M. Arago à l'Académie des sciences au nom de la commission qui avait été chargé d'examiner les mémoires envoyés au concours pour le prix de la diffraction [read on March 15, 1819]," *Oeuvres de Fresnel* 1:229; henceforth referred to as "Arago's 1819 report."

86 A. Fresnel to L. Fresnel, "September 5, 1818," *Oeuvres* 2:842.

87 Verdet, "Introduction," LXXXVII.

de France.[88] He also favorably referred to Fresnel's version of the principle of interference.[89]

Poisson admired Fresnel's observations of diffraction and agreed that the emission theory could not explain them.[90] In 1818, he began his research on the theory of reflection and refraction of longitudinal waves at the boundary of two media, which he believed to be valid for both sound and light. He obtained the correct formula for the intensity of reflected light at normal incidence.[91] However, his 1823 attempt to extend the theory to oblique incidence was unsuccessful, for he found two Brewster's angles for non-polarized light.[92] This failure, nonetheless, did not prompt Poisson to abandon his wave theory, and in 1831 he corrected it. He did not object to the usefulness of the Huygens-Fresnel principle, but he did not see any advantage in it over the wave equation. He criticized Fresnel's theory for lack of mathematical rigor and mechanical foundation and insisted that if a wave theory can be constructed, it must be based on the wave equation. Contrary to other "emissionists," he accepted the principle of interference as an undulatory concept, treating it as an extension of Bernoulli's principle.[93]

While examining Fresnel's contest memoir Poisson found that Fresnel's integral can be easily calculated when a diffracting body is a circular screen or a circular aperture, and that the screen must produce an unusual phenomenon – a bright spot at the center of its shadow.[94] Arago's experiment confirmed Poisson's prediction, which, according to some historians, decided the fate of the prize and considerably accelerated the acceptance of the wave theory of light.[95] In my view, both claims are exaggerated, for neither a few lines on this episode in Arago's

---

88 Fresnel, *Oeuvres* 1: 183, 201, 217, 533.

89 Biot, "Mémoire sur les rotations imprimées par certain corps," (see n.33, above). Biot gave a brief account of Young's ideas on chromatic polarization and of some of his experimental results and emphasized the agreement between Fresnel's theory and his own observations. Fresnel mentioned ". . . all the compliments that I could have received from MM. Arago, Laplace, or Biot . . ." (Fresnel to Young, November 26, 1824, *Oeuvres* 2: 771). Thus, encouragement from Laplace and Biot was not unusual fo Fresnel.

90 "Extrait d'une lettre de M. Poisson à M. A. Fresnel, March 1823," *Annales de Chimie* 22 (1823), also *Oeuvres* 2: 213; A. Fresnel to L. Fresnel, September 5, 1818, ibid., 849; and A. Fresnel to L. Fresnel, November 15, 1818," ibid., 852.

91 Poisson, "Sur le mouvement des fluides élastiques dans des tuyaux cylindriques, et sur la theorie des instruments à vent," *Mem. Acad.* 2 (1817): 380–81.

92 Poisson, "Extrait d'un memoire sur la propagation du mouvement dans les fluides élastiques [1823]," *Ouevres de Fresnel* 2: 202–5.

93 "Extrait d'une lettre de M. Poisson à M. A. Fresnel," 207, 213; and "Lettre de M. Poisson à A. Fresnel, [March 6, 1823]," ibid, 187.

94 Arago, "1819 Report," 236, and Fresnel, "Mémoire courronné. Note I," 365. This note was published in 1826, but part of it was reported by Arago as "Note (E)" to his "1819 Report," 245–6.

95 Whittaker, *History*, 108; and Silliman, "Augustin Fresnel," 198. For a typical for physicis textbooks view see in S. G. Lipson & H. Lipson, *Optical Physics*, 2nd ed. (Cambridge, 1981), 4.

report nor the comments on it in the 1820's show any particular excitement. The reason why the "undulationists" did not gain much from this allegedly remarkable confirmation of Fresnel's theory is evident: until 1826 Arago's brief comment was the sole source of information about the "Poisson's spot."

As to its role in the decision of the contest commission, it seems that two factors of importance to this problem have not yet been discussed. First, although Arago is speaking of "Poisson's spot" as a "singular" phenomenon, its novelty was relative, since it resembled Maraldi's observation of a bright spot at the center of a sphere's shadow, and Biot had cited Maraldi's paper.[96] Second, Arago's report shows that Poisson's result was viewed by the commission as challenging not the wave theory of light – as it is generally believed – but only one aspect of Fresnel's theory of diffraction related to the intensity of fringes. The concept of intensity of fringes was Fresnel's major contribution to the theory of diffraction, but he did not verify quantitatively his formulae for intensity. Arago was concerned with photometrical problems and considered this aspect of Fresnel's theory to be very important. He treated his experiment with a circular screen as an indirect test of the formulae for intensity. In fact, since this observation was qualitative, it could not prove anything. Apparently, Arago realized this, for he urged Fresnel to confirm his theory by direct measurements of intensity. Fresnel accomplished this task for diffraction by a circular aperture. He calculated the positions of an observation screen, given the aperture's diameter and its distance from the luminous source, necessary to observe the central bright spot of a particular color. Fresnel's observations fully confirmed his theoretical predictions, and Arago immediately published a short account of this work in a note to his report.[97]

Two points must be emphasized here. First, Poisson suggested to Fresnel to study diffraction by a circular screen and a circular aperture twice: before the prize was awarded and after that, and Fresnel performed his investigation of the circular aperture *after* he received the prize. Second, Arago delayed the publication of his report by more than three months. Taking into account the importance of this publication to Fresnel, I suppose that Arago was waiting for the completion of Fresnel's work on the circular aperture, for he believed that it will provide a strong support for his theory. Since neither Arago nor Poisson objected to a continuation of Fresnel's research on diffraction by a circular aperture – the only conclusive case of the two suggested by Poisson – after the contest, they did not connect this work with the contest. Therefore, the commission decided that Fresnel's memoir deserved the award *despite* the lack of experimental support for his

96 Biot, *Traité de physique* 4:774.

97 *Annales de Chimie* 11 (May 1819): 1–17 (report) and 18–30 (notes), see especially pp. 29–30. This issue appeared early in July.

formulae for intensity of fringes, and Poisson's discovery played no significant role in adjudging the prize to Fresnel.[98]

The grounds for the commission's decision would be clearer if Fresnel's memoir could be compared with his rival's. The latter has not been preserved in the Academy's archives, but there is, in my opinion, sufficient indirect evidence to identify its author and the principal content of his memoir. I supposed that the unknown author had had some experience in the study of diffraction before entering the contest, and that he somehow utilized his work after it was rejected by the commission. According to these criteria, I examined papers on diffraction that appeared between 1804 and 1820. Starting with ten candidates, I gradually excluded all but one. First, I eliminated those who presented only qualitative observations or considerations (Mollweide, W. Herschel, C. A. Prieur de la Côte-d'Or (1763–1832), Georg Friedrich Parrot (1767–1852), Professor in Dorpat, Heinrich Wilhelm Brandes (1777–1834), Professor of Physics in Leipzig).[99] Venturi's work was mathematical and supported by measurements, but his topic, supernumerary rainbows, was too narrow.[100] Brewster published a summary of his earlier observations on diffraction in the summer of 1817, and Johann Tobias Mayer, Jr. (1752–1830) read a paper on diffraction on August 1, 1818.[101] These authors must be excluded, since they made their results and names public before the contest was over.[102] Pouillet was probably considered a potential candidate by the academicians. However, he obviously was not, since, according to Arago's report, the author of the other memoir was familiar neither with Young's works nor with Fresnel's 1816 memoir on diffraction.

The only possibility with no obvious objections is Honoré Flaugergues (1755–1830), a French astronomer and physicist, and a corresponding member of the Paris Academy of Sciences. He won several prizes from provincial acade-

---

[98] Verdet remarked that the experimental confirmation of "Poisson's" spot made a "great impression" on the judges, however, he belived that this affected only "unanimity" of their decision; see "Introduction," XLII. Worrall's conclusion about the role of "Poisson's spot" is similar to mine, although some of his arguments are different (see his "Fresnel, Poisson and the white spot..." in *The User of Experiment*, ed. by D. Gooding et al., (Cambridge University Press: Cambridge, 1989), 135–157.

[99] See n. 110, ch. VI.

[100] Venturi, *Commentari sopra la storia e le theorie dell' ottica* (Bologna, 1814). See Brandes' comment on this book in *Annalen der Physik* 52 (1816): 385–97.

[101] On Brewster see n. 69, above. Mayer, "Phaenomenorum ab inflexione luminis pendentium ex propriis observationibus et experimentis recensio et comparato," *Göttingen Commentationes* 4 (1820): 49–80. An abridged account was published in *Journal für Chemie u. Physik* 25 (1819): 231–37.

[102] To guarantee the objectivity of the commission's judgement, memoirs presented to the contest instead of the author's name had a motto which was kept in a sealed evenlope until the commission made its decision. Any public presentation of a memoir under the author's own name (as with Brewster and Mayer) would have revealed his identity to the commission.

mies in France for optical subjects, including a prize for the best experimental work on diffraction of light in 1811. His memoir contained a large number of observations, mostly qualitative, which demonstrated that diffraction fringes were unaffected by variations of a body's material, shape (provided its cross-section was preserved), temperature, state of electrification or magnetization, and of the index of refraction of the surrounding medium.[103] It would have been natural for Flaugergues to prepare for the new contest by improving his technique, increasing the number and precision of his measurements, deriving some mathematical relations between them, and extending his observations to some new phenomena, according to the requirements of the contest. And these are precisely the additions he made to his old memoir, according to the paper which he published in September 1819.[104] Since Flaugergues is the only physicist to publish on diffraction both before and after the contest, with improvements correspoding to its program, I believe that it is most likely that he is the one who competed with Fresnel.[105]

In his 1819 paper Flaugergues said that he made a number of measurements of the fringes' positions with a mat glass screen and a lens focused on it. He determined his precision to be 0.02 mm. Most of his results were not connected by mathematical laws, except for the law that the distance of a fringe from the center is proportional to the index of refraction of the surrounding medium (it seems that he continued the work of Biot and Pouillet who discovered this law for water only). Flaugergues was primarily concerned with qualitative study of various factors that influence the diffraction fringes. In addition to the observations that he had described earlier, he also studied whether the positions of the fringes depend on the light source (the sun, stars, planets, flames), on reflection or refraction of light prior to diffraction, and on the polarization of light. Although Flaugergues was an "emissionist," the commission did not find that his work satisfied its standards. The report stated that his observations were not very precise, that he overlooked some phenomena when dealing with diffraction by a small aperture, and that he was not familiar with works of Young and Fresnel, and consequently,

103 See n. 100, ch. V.

104 Flaugergues, "Supplément à différent mémoires sur la diffraction de la lumière, publies dans le 'Journal des Physique'," *Journal de Physique* 89 (September 1819): 161–86.

105 The only other author who published on diffraction between 1819 and 1829 was Fraunhofer, but he explored a very narrow field (the diffraction grating), and such a paper could not have been presented to the contest.

"the part of his work, which relates to the *influence that intersecting light rays exert on one another*, far from adding anything to that what is already known, contains many evident errors."[106]

This is an unequivocal recognition of the principle of interference as an indispensable means for the study of diffraction. Hence, it took two years (from March 1817 to March 1819) for the Paris Academy of Sciences to pass from an implicit and indirect recognition of the principle of interference to an explicit and direct one. Apparently, Fresnel's prize-winning memoir played a significant role in this transition. I will now examine individual reactions of scientists to this work.

VIII.4 Response to Fresnel's prize-winning memoir

The first public support of Fresnel came in 1819, when Biot included a ten-page account of the principle of interference and its applications in his optical supplement to the translation of Ernst Fischer's physics textbook.[107] Therefore, as early as 1819 French students were able to study this most advanced and difficult concept in optics. At the same time, Biot taught the principle of interference and Huygens-Fresnel principle in the course of physics at the *College de France*.

The next response came from Scotland. In 1820, Brewster published an English paraphrase of Arago's 1819 report with several comments of his own.[108] He called Fresnel a "philosopher of the highest promise," whose discoveries respecting diffraction are "in the highest degree important, and cannot fail to be regarded as affording a strong confirmation of the Huygenian theory of light."[109] Earlier, in 1816 Brewster considered Young's theory of double refraction to be "an argument in favour of the undulatory system."[110] Thus Brewster, like Biot, but even more enthusiastically, praised the achievements of his opponents without changing his own views. As shown in Ch. VI, Brewster first acknowledged the principle of interference in 1815; and in 1817 he showed even more enthusiasm

[106] Arago, "1819 report," 237, italics added.
[107] See n. 25, above.
[108] Brewster, "Account on Fresnel's discoveries respecting the inflexion of light," *The Edinburgh Philosophical Journal* 2 (1820): 150–53.
[109] Ibid, 150.
[110] Brewster, "On the communication of the structure of doubly refracting crystals to glass . . . and other substances, by mechanical compression and dilatation," *Phil. Trans.*, 1816: 178.

for Young's "beautiful law of interference."[111] This does not mean that in 1817 he grasped this concept better than in 1815, for Young again found Brewster's suggestion for a new application of the principle of interference to be wrong.[112]

The first applications of the principle of interference by other authors than Young and Fresnel appeared in 1823. Poisson noticed that Young's and Fresnel's theory of the colors of thin films could not explain the perfect darkness of the minima in reflected light. Instead of rejecting the principle of interference, he proposed to improve it by taking into account more than two interfering rays, namely, those produced by multiple reflections within the plate.[113] Joseph Fraunhofer, un "undulationist", applied the principle of interference to derive a mathematical law for bright fringes produced by a diffraction grating.[114] He found that

$$\sin \Theta_m = \sin \Theta_o \pm m \frac{\lambda}{d}; \quad m = 0, 1, 2, \ldots \tag{8.1}$$

where Θ_o is the angle of incidence, Θ_m the angle of diffraction of the mth fringe, d the grating's constant, and λ the wavelength. As mentioned in Ch. VI, Fraunhofer based his theory on Young's two-ray model of interference. He heard something about Fresnel's work, but neither the extent of his knowledge nor his source of information are known.[115]

Some comments on the principle of interference that appeared after 1821 could have been influenced not only by Fresnel's 1819 memoir on diffraction but also by his 1821 paper on chromatic polarization. After a long delay, on June 4, 1821, Arago at last presented his report on Fresnel's theory of chromatic polarization. He was primarily concerned with a demonstration that Fresnel's theory, based on the principle of interference, refuted Biot's theory of "mobile polarization." In his reply Biot argued that his theory explained the phenomena as well as Fresnel's, and perhaps even better, for it did not include any hypothetical elements. The exchange between Arago and Biot was much too personal, and while the Academy recommended Fresnel's work for publication, it refused to include

---

111 Brewster, "On the laws of polarisation and double refraction in regularly crystallized bodies [written July 1, 1817]," *Phil. Trans.*, 1818: 271–72.

112 Ibid., 273.

113 Poisson, "Note sur le phénomène des anneaux colores," *Oeuvres de Fresnel* 2: 229–46.

114 See n.99, Ch. V. Fraunhofer's paper was presented to the Munich Academy of Sciences on July 14, 1821. The principle of interference is repeatedly mentioned there, as well as the author's sympathy for the wave theory. There is a very clear remark (p. 111) that Fraunhofer either already had his equation or knew how to derive it. For some unknown reason he published this equation only two years later, in his "Kurzer Bericht." Incidentally, Fraunhofer believed that the dark lines in the solar spectrum resulted from interference, an idea, which became quite popular in Germany.

115 Fraunhofer, "Kurzer Bericht," 359.

in its minutes either Arago's report or Biot's answer, thus showing an unwillingness to decide between the competing theories.[116] Likewise, Fourier does not mention this debate in the official analysis of the Academy's achievements for 1821–22. Instead, he praises Fresnel, who "has determined mathematical laws of the most complex phenomena, and all the results of his analysis exactly correspond to observations," and the principle of interference as the "most far-reaching and fruitful notion of this new optics." [117] Nothing is said about the wave origin of this principle. In other place Fourier said that the object of Fresnel's most recent researches is the "*mathematical expression* of laws" of double refraction, reflection, and other phenomena,[118] but there is no word about transverse light waves. A similar attitude is seen in all reports on Fresnel presented after 1816. If we compare the 1819 report on his memoir on diffraction (the majority of reviewers were the "emissionists") with the 1821 report on his work on chromatic polarization and the 1822 report on his memoir on double refraction (both were reviewed by the "undulationists" Arago, Ampere, and Fourier), we find no difference in style. All three praised Fresnel's experimental skill and his ability to derive mathematical laws which explain a variety of phenomena, and none of them said anything about the wave theory of light.

When summarizing these reviews one may conclude that they differed very much from the early reviews of Young in two points. One is that between 1816 and 1822, the principle of interference won an approval of the leading scientists of the time, most of whom were, incidentally, the "emissionists." Another is that no one mentioned Fresnel's failure to convert him to the undulatory doctrine or to answer Newton's objections to it. This is puzzling because both statements would have been correct: of all reviewers Arago was the only true convert for that period, and like Young, Fresnel did *not* give an undulatory explanation of the rectilinear propagation of light (although, unlike Young, he held it in his hands). Both differences can be explained by the fact that, unlike Young's case, the reviewers of Fresnel's works were all "quantitative physicists." In their view, a discovery of a mathematical law of phenomena was more important than a support for a particular hypothesis of the nature of light. Thus, they greeted the principle of interference of light as a means to quantify the theory of periodical colors even if they denied its undulatory implications. By 1818, the "quanitative physicists"

[116] On this controversy see *Oeuvres de Fresnel* 1: 533–600, especially 553 n; also see Buchwald, *Wave Theory*, 239–244.

[117] Fourier, "Analyse des travaux de l'Académie royal des sciences, pendant l'annee 1822. Partie mathématique, [read on April 24, 1823]," *Mém. Acad.* 5 (1821–22): 235.

[118] Ibid. Compare this with the report by Arago, Ampère, and Fourier on August 19, 1822 on Fresnel's memoir on double refraction. Granting the author full credit for mathematical solution and experimental verification of such a difficult problem, the commission declined to discuss the undulatory concepts that were the basis of the theory (see *Procès-verbaux)* 7:359–61).

set the fashion of doing science in Paris, and to them pursuing a problem such as "why does light travel rectilinearly?" had to appear a waste of time because it seemed to be unreducible to a mathematical law.

I neglect here the opinions of the "qualitative physicists" and "non-experts", since my primary concern is with the *development* of the principle of interference, which could have come only from the "quantitative physicists." From their prospective, by 1822 the principle of interference of light was firmly established, and my next task will be to investigate how the adoption of this concept affected the dispute on the nature of light.

VIII.5 Principle of interference and the wave theory

Between 1823 and 1825 several books appeared in France and Germany which defended the wave theory and treated the principle of interference as an integral part of it. Their authors were Eugene Péclet (1793–1857), a professor at the *College Royal de Marseille*; Andreas Baumgartner (1793–1872), a professor at the University of Vienna; Jacques Babinet (1794–1872), a professor at the *Lycee de Saint-Louis* in Paris; César Despretz (1789–1863), a professor at the *Collège Royal de Henry IV* in Paris; Ernst Weber (1795–1878) and Wilhelm Weber (1804–1891), professors in Leipzig and Halle, respectively; and Alexandre Bertrand (1795–1831), a physician and popular science writer.[119] The French authors relied primarily on Fresnel's "De la lumière," while the Germans, not yet familiar with it, referred to Fresnel's earlier works. The article "De la lumière" itself also became an important source for the dissemination of the principle of interference and the wave theory. In January 1823, extracts from this paper appeared in the *Bibliothèque Universelle* in Switzerland.[120] In 1825, Johann Poggendorff (1796–1877) began to publish its German translation in the *Annalen der Physik*.[121] In Britain, Brewster started publishing Young's translation of this article in

[119] Péclet, *Course de physique* (Marseille, 1823), 487–99); Baumgartner, *Die Naturlehre nach ihrem gegenwärtigen Zustande mit Rücksicht auf mathematische Begründung*, 3 vols (Wien, 1824), 2: 91–96, 101–2, 108–119; Despretz, *Traité élémentaire de physique* (Paris, 1825), 603–30; Babinet and Charles Bailly, *Resumé complet de la physique des corps imponderable*, (Paris, 1825), 152–211. On the brothers Weber see n.21 (ch.I); Bertrand, *Lettres sur la physique*, 2 vols. (Paris, 1824–25), 2: 303–307.

[120] Fresnel, "De la lumière (extrait)," *Bibliothèque Universelle* 22 (1823): 3–17.

[121] Fresnel, "Ueber das Licht," *Annalen der Physik* 3 (1825): 89–128, 303–28; 5 (1825): 223–56; 12 (1828): 197–249, 366–99.

1827.[122] In the preceding decade the only account of the principle of interference in English (besides Young's 1817 article "Chromatics") was that of Biot from his 1819 "Optical Supplement" to Fischer's book, translated by Henry Coddington.[123] In 1827, John Herschel gave the first mathematical account of Fresnel's theory in his article "Light," whose first part (unpolarized light) was translated into French in 1829.[124] These works plus Pouillet's 1828 *Lectures* seem to exhaust all of the secondary sources on the wave theory of light before 1830.[125]

It is important to note that Fresnel's hypothesis of transverse light waves did not attract much interest for several years. It was only briefly mentioned by Péclet, Babinet, and Despretz. The *Bibliothèque Universelle* completely omitted that part of the "De la lumière" which dealt with this concept and its application, and the *Annalen der Physik* postponed the publication of this part until 1828. This hypothesis and its application to double refraction and chromatic polarization became widely known to scientists only after 1827–1828.[126] Ampere's 1828 paper was the only new application of transverse light waves before 1830. Therefore, *in the 1820's the wave theory of light was reduced to the theory of interference*. And even in this narrow field not much creative work was done, for after Fraunhofer's and Poisson's works mentioned above only Herschel and Babinet applied the principle of interference.[127] Most authors did not try to go beyond explaining the principal idea of interference and some of its simplest applications.

Like the concept of transverse light waves, the Huygens–Fresnel principle did not receive much prominence in the 1820's. Biot, as mentioned above, taught this principle in his physics course since 1819, and he also described it briefly in his textbooks.[128] In 1819, Young admitted the Huygens–Fresnel principle as a

---

[122] Fresnel, "Elementary view of the undulatory theory of light," *Quarterly Journal of Science* 1 (1827): 127–41, 441–54; 2: 113–35; 3 (1828): 198–215; 4: 168–91, 389–407; 5 (1829): 156–65.

[123] Coddington, *An Elementary Treatise on Optics* (Cambridge, 1823), 152–162.

[124] Herschel, *Traité de la lumière* vol. 1 (Paris, 1829).

[125] Pouillet, *Leçons des physique de la Faculté des Sciences de Paris* 2 vols (Paris, 1828), see vol. 2. I have been unable to consult this book recently and so cannot specify the pages. There were also popular accounts of Fresnel's works in articles. See Baumgartner, "Die circlare Polarisation des Lichtes nach Fresnel's Arbeiten," *Zeitschrift für Physik* 2 (1827): 1–20, "Ein Beitrag zur Theorie der Beugung des Lichtes," ibid, 3 (1827): 443–51; and Karl Marx, "Ueber das lichtbrechende Vermögen der Körper," *Journal für Chemie u. Physik* 52 (1828): 385–411.

[126] Particularly important was the publication of Fresnel's "Mémoire sur la doble refraction," (*Mém. Acad.* 7 (1827): 45–176) and Herschel's article "On light."

[127] On Herschel's account of the colors of thick plates see Ch. IV. In 1827 Babinet presented to the Philomatic society a memoir on periodical colors. The part on the colors of a grating was published simultaneously with Young's paper on the same subject (see "Sur les couleurs des reseaux," *Annales de Chimie* 40 (1829): 166–77), while other parts were published later as "Mémoire sur les couleurs des doubles surfaces à distance," *Comptes Rendus* 7 (1838): 694–98. Both Herschel and Babinet applied the two-ray interference model.

[128] Biot, "Additions à l'optique," 418–19, and *Précis élémentaire de physique expérimentale*, 2 vols., 3rd ed. (Paris, 1824), 2: 471–73.

mathematical law but considered it mechanically impossible.[129] Poisson expressed similar views in 1823, though his criticism was much stronger and more specific.[130] Fresnel defended his theory with much ability, yet neither persuaded the other. Although hardly anyone at the time properly understood the essence of Poisson's criticism, the mere fact that this famous mathematical physicist objected to Fresnel – a fact noted by many authors – hindered the acceptance of the Huygens-Fresnel principle and the theory of interference.[131] Around 1830 Paul Fuss (1797–1855), Secretary of St. Petersberg Academy of Sciences, and John Barton (1771–1834), Comptroller of the Mint, brought forth new objections to the Huygens-Fresnel principle.[132] Besides its mechanical improbability, the mathematical complexity of the principle was another reason for its unpopularity. Physics textbooks, even by "undulationists," omitted it altogether or described it very briefly and mostly qualitatively. The best French account, though purely qualitative, was Péclet's.[133] Baumgarten sketched the mathematical aspect of the Huygens-Fresnel principle, but only as much as was necessary to understand the calculation of the path difference of interfering rays, but not their intensity.[134] Herschel was the only author to present a full mathematical account of this principle, but neither he nor anyone else tried to clarify Poisson's objections. More importantly, in the 1820s, no one attempted to apply the Huygens-Fresnel principle, and one may say that at the time its role was insignificant. It was viewed just as a peculiar way of applying the principle of interference, with many rays originating at various distances from the diffractor instead of two rays coming from its edges.

Although all physicists were impressed by the close agreement of Fresnel's theoretical predictions with his measurements, there is no evidence that anyone demanded that the wave theory not be accepted unless *all* the explanations based on it reached the same precision as his theory of diffraction. On the contrary, the reviewers of Fresnel's papers agreed with the author that his explanations of chromatic polarization, reflection, and double refraction were in excellent agreement with observation (only some of the measurements presented by Fresnel were his own), although it was no better than that of Fresnel's 1816 the-

129 Young to Arago, August 4, 1819, *Oeuvres de Fresnel* 2: 746.

130 On the Poisson-Fresnel controversy see *Oeuvres* 2: 183–238, and Buchwald, *Wave Theory*, 188–198.

131 See, for instance, Biot, *Précis de Physique*, 3rd. ed., 2: 472; Webers, *Wellenlehre*, 559; and Brandes, "Inflexion des Lichtes," *Physikalisches Wörterbuch* 11 vols. (Leipzig, 1825–45), 5: 724 n.

132 Barton, "On the inflection of light, in reply to Professor Powell," *The Philosophical Magazine* 3 (1833): 176–78; and Fuss, "Sur le prix de physique proposée par l'Académie imp. des sciences de St. Petersbourg," *Bulletin des Sciences Mathématique* (1828): 98, 102.

133 Peclet, *Course de physique*, 492–96.

134 Baumgartner, *Die Naturlehre* 2: 111–14.

ory of diffraction whose "extreme precision" Biot praised in 1817. This shows that the high precision achieved by Fresnel in his 1818 diffraction experiments *exceeded* the requirements of contemporary optics and explains why physicists who were studying diffraction did not have any stimulus to abandon the simple two-ray model in favor of the Huygens-Fresnel principle. Hence, for several reasons the Huygens-Fresnel principle did not assist much in the establishment of the theory of interference in the 1820's, and this theory was viewed as essentially the same as Young's. This fact justifies the possibility of comparing the response to Young's and Fresnel's theories even after 1819.

Let us now compare the reception of the principle of interference and that of the wave theory between 1816 and 1830. Among scientific bodies the Paris Academy of Sciences was the first to approve the principle of interference in 1816, and the Royal Society of London was the first to recognize the wave theory when in 1827 it awarded Fresnel the Rumford medal for the undulatory theory of polarization. Physics textbooks and university courses began to explain the principle of interference in 1819, while the wave theory initially appeared in a textbook in 1823. The first application of interference was in 1816, whereas the first development of the wave theory occured in 1819. Thus by any of these three criteria the adoption of the principle of interference *preceded* that of the wave theory. It is necessary to note that "quantitative physicists" made the most important contributions to the acceptance of both the principle of interference and the wave theory, and that the "emissionists" among them, particularly Poisson, played a significant role. Thus we must conclude that the situation in physical optics in the early nineteenth century can hardly be understood in the traditional terms of a struggle between the "emissionists" and "undulationists" over the true theory of light.

Arago said that although only a few people appreciated Young's work, "in such matters it is wiser to *weigh* votes than to count them."[134a] He certainly referred not only to Young but to the whole situation with the wave theory after 1815. Following Arago, I believe that the development of the wave theory depended on those few who were able to *apply* the new ideas rather than on a large (comparatively) number of those who called themselves "undulationsts" but did nothing for advancing the theory. One may argue, that the latter "undulationists" might have served the wave cause through their teaching. This however, seems not to be the case with the scientists most active in the wave theory of light in the 1830's. Apparently, most of them learned the wave theory directly from Fresnel's works supplemented by Herschel's article and Airy's book and not from the text-

[134a] Arago, "Thomas Young", *Oeuvres*, v. I, 258.

books or lectures.[135] It is necessary to mention here that although the wave theory of light received support from Poisson immediately after 1816, it was far from being popular at the time, despite some statements of contemporaries to the contrary. In 1820, the physician Alexander Marcet, wrote that "the poor 'emissionists' lose ground every day."[136] A month later the chemist Jacob Berzelius (1779–1848) asserted that, "currently, in Paris Euler's hypothesis of light is in favor, and those who share Newton's opinions would be certainly declared fools."[137] And a few months after that the zoologist Henri-Marie Ducrotay de Blainville (1777–1850), the editor of the *Journal de Physique*, wrote that

> the emission theory apparently must yield to the undulatory theory, which is *quite generally adopted in England,* as is proved by *many articles* published on this subject in the journals of this country.[138]

This statement is patently false, and so probably is Berzelius's, unless both of them identified all criticism of Newton's views with a support of the wave theory. This shows, just as in Young's case, that the opinions of the unknowledgeable can be misleading if one attempts to draw conclusions from them as to the general attitude toward the principle of interference or the wave theory of light. Many Frenchmen who called themselves "undulationists" were probably influenced not by Fresnel's memoir itself, but rather by the Academy's prize awarded to it in conjunction with Arago's criticism of the emission theory. However large their number, the wave theory itself gained little from them, since they did not understand it. They could have only indirectly affected the establishment of the wave theory, since it was probably easier for Fresnel to work in an atmosphere favorable to his theory, even if it was misunderstood, than it was for Young in his time.

---

135 These scientists were: George Airy (1801–1892), Babinet (1794–1872), Cauchy (1789–1857), George Green (1793–1843), William Hamilton (1805–1865), Philip Kelland (1808–1879), Karl Knochenhauer (1805–1875), Humphrey Lloyd (1800–1881), Franz Neumann (1798–1895), Powell (1796–1860), and Friedrich Schwerd (1792–1871). By the early 1820's some of them, such as Cauchy, Babinet, Neumann, Schwerd, and Powell had completed their education. Lloyd, Hamilton, and Airy were students at this time, but neither in Cambridge nor in Dublin was the undulatory theory then taught. Kelland and Green, who graduated from Cambridge in 1834 and 1837, respectively, were influenced by the second generation of "undulationists.' Nothing is known about Knochenhauer's education.

136 "Marcet to Berzelius, September 15, 1820," *Jac. Berzelius Brev, Supplement 2,* vol. 1, pt. 1 (Stockholm, 1941), 211.

137 "Berzelius to Marcet, October 19, 1820," ibid, 211.

138 *Journal de Physique* 90 (1819): 40–41.

VIII.6 Principle of interference and the emission theory

Not only "undulationists" popularized the principle of interference, but so did some of the "emissionists." Haüy and François Beudant (1787–1852), Professor of Mineralogy at the *Faculté des Sciences de Paris*, included an account of interference in their textbooks.[139] Although they acknowledged the inability of the emission theory to explain interference, they still preferred it to the wave theory. Biot asserted that there was no contradiction between the emission theory and the principle of interference. His view, presented in the supplement to Fischer's book in 1819, was based on an hypothesis of Young published in his "Chromatics" in 1817.[140]

By 1817 Young had become pessimistic about the possibility of constructing in the near future a wave theory that could explain all optical phenomena. He decided to focus on his principle of interference as "the only practical mode of connecting an immense variety of facts with each other." [141] Consequently, Young formulated this principle as a simple generalization of facts, without invoking the wave theory.[142] Though he did not abandon this theory, he discussed the possibility of adjusting the principle of interference to the emission theory:

> if we suppose, with Newton, the projected corpuscles of light to excite sensation by means of the vibrations of the fibres of the retina and of the nerves, we may imagine that such vibrations must be most easily produced by a series of particles following each other at equal distances, each color having its appropriate distance in any given medium: it will then be demonstrable, that any second series of similar particles, interfering with them, in such a manner as to bisect their intervals, will destroy their effect in exciting a vibratory motion; each succeeding particle meeting the fibre at the instant of its return from the excursion occasioned by the stroke of the preceding, and thus annihilating the motive effect of that stroke.[143]

Young indicated a defect in this model: the principle of interference requires light to travel faster in a rarer medium, which contradicts one of the fundamental

[139] Haüy, *Traité élémentaire de physique*, 2 vols, 3rd ed. (Paris, 1821), 2: 381–83, and Beudant, *Essai d'un cours élémentaire et général des sciences physiques. Partie physique*. 3rd ed. (Paris, 1824), 556.

[140] Biot, "Additions à l'optique," 412–14; see also his *Précis élémentaire de physique expérimentale*, 2 vols (Paris, 1824), 2: 459–60.

[141] Young, "Chromatics," 279.

[142] Ibid., 287.

[143] Ibid., 328.

results of the emission theory. Fresnel noted that the model could not explain the coherence of the motion of luminous particles and the existence in a luminous body of vibrations of different frequency responsible for the emission of light of different colors.[144]

Biot's chief idea is that,

> "one can, without a violation of any rule of logic, equally comprehend the principle of interference in the emission system, by making the result which it expresses the *condition of visibility*."[145]

He assumes that the sensation of brightness appears when all rays enter the eye in the same phase, and the sensation of darkness when they enter in the opposite phase and destroy each other. Therefore, according to Biot, interference occurs inside, not outside, the eye. He does not, however, specify the mechanism for such an interference. Biot then makes an important remark that two rays coming to the eye from a white screen (or a semitransparent glass) employed for observing fringes maintain in the eye the same relation of phases of their fits that they had at the screen. Hence, the only thing that one needs to know for calculating the intensity and positions of fringes is the path difference of rays *at the screen*. This is exactly the same requirement that is applied in the wave theory, so that all the mathematical apparatus of the wave theory can be utilized in the emission theory. Thus, Biot demonstrates that the concepts of interference and periodicity can be made compatible with the emission theory provided one abandons attempts to build their mechanical models.

There was, however, one more obstacle to overcome. As mentioned above, Newton argued that light propagates faster in a denser medium, while the principle of interference requires the contrary. To retain the principle of interference, Biot decides to sacrifice Newton's condition. He proposes to treat the principle of interference as an experimental law, following Young's definition in "Chromatics," which postulates that the velocity of light is inversely proportional to the index of refraction of the medium.[146]

In his attempt to develop the emission theory of interference Biot acquired several followers. Some, such as Brewster and Fuss, adopted his physiological model of interference.[147] Others, as Potter and Barton, tried to construct a me-

---

[144] Fresnel, "Mémoire couronné," 259–61.

[145] Biot, "Additions à l'optique," 412

[146] Ibid.

[147] Brewster, "On the undulations excited in the retina by the action of luminous points and lines," *The Philosophical Magazine* 1 (1832): 170; and Fuss, "Sur la prix de physique," 107.

chanical model of the periodicity of light.[148] Richard Potter (1799–1886), later Professor of Physics at the University College of London, deduced from his observations that light propagates slower in a denser medium.[149] However, he did not think that all consequences of the emission theory of interference must coincide with the corresponding results of the undulatory theory. The latter demands, for instance, that the central bright fringe occurs where the path difference of interfering rays is equal to zero. Potter observed, however, that "Fresnel's mirrors" gave him a *dark* fringe where he expected a zero path difference.[150] He therefore modified the principle of interference in such a way that the maxima and minima corresponded to the path differences equal to the odd and even number of the half-wavelengths, respectively. While adopting the theory of interference as a phenomenological theory, Potter concluded that his experiments showed the fallacy of the wave theory of light.

One of the reasons for objecting to the undulatory model of interference was a miscomprehension of energy transformations in interference phenomena. In 1820, Berzelius stated that the wave interpretation of interference is unnecessary, for the mutual destruction of two waves is equivalent to the destruction of the motion of two identical colliding bodies moving in opposite directions. Since in the latter case, he continued, the motion is transformed into heat, it is possible that the formation of a dark fringe is accompanied by an increase in temperature.[151] In 1828, Fuss judged Fresnel's account of Arago's observation of interference of ultraviolet light to be erroneous. He ignored Fresnel's assumption that the chemical effect of light is proportional to the maximal velocity of vibration of particles and instead supposed it to be in direct relation to the amount of absorbed light. Then he concluded that it is impossible that in some points two light beams would produce a smaller chemical effect than either of them alone.[152] Evidently, when thinking of particles of light, as Berzelius and Fuss did, it is difficult to grasp the undulatory model of the energy transfer. However, in 1836 the same thing happened to Augustin Cauchy (1789–1857), a leading French "undulationist," who insisted that interference of light violates the conservation of energy.[153]

---

148 Potter, "On the modification of the interference of two pencils of homogeneous light . . .," *The Philosophical Magazine* 1 (1832): 83; and Barton, "On the inflection of light, in reply . . .," 177–78.

149 Potter, "A reply to the remarks of Professors Airy and Hamilton on the paper upon the interference of light," *The Philosophical Magazine* 2 (1833): 281, and "Particulars of a series of experiments and calculations undertaken with a view to determine the velocity with wich light traverses transparent media," ibid., 340: 42.

150 Potter, "On the modification of the interference," 83–84.

151 "Berzelius to Marcet, October 19, 1820," 8.

152 Fuss, "Sur le prix de physique," 106–107.

153 Cauchy, "Notes sur l'optique," *Comptes Rendus* 2 (1836): 349.

Had Fresnel published his explanation of the redistribution of energy due to interference (see VII.2), such miscomprehensions would perhaps not have occured.

Barton denied the undulatory nature of interference, for, in his view, it requires that light waves meeting in a small opening affect one another.[154] According to him, this contradicts the whole theory of Huygens and also our experience of distinct vision through a small hole or in a refracting telescope where the rays repeatedly cross when passing through the instrument. Barton's arguments are similar to those of the eighteenth century (Ch. II), which shows that even after three decades of the application of the concept of interference its physical meaning was still obscure.

VIII.7 Understanding of coherence after Fresnel

One can argue that Barton's miscomprehension of the concept of interference is not surprising, since he opposed the wave theory and consequently was uninterested in such subtleties as coherence. It is natural to suggest that the "undulationists" had to fully grasp the concept of coherence in order to be able to counter the attacks of the "emissionists;" and that they could have easily accomplished this task by reading Fresnel's 1822 article "De la lumière." The evidence shows, however, that this was not the case.

The initial accounts of the principle of interference by the French "emissionists" Biot, Haüy, and Beudant omitted this concept altogether. The first French followers of Fresnel, on the contrary, discussed coherence in some detail, though without giving a complete list of the conditions of coherence. Despretz, for instance, mentioned the condition of direction and explained the necessity for a common source.[155] Babinet clarified the requirement for a small size for the luminous source and the relation between the degree of monochromaticity of light and the number of visible fringes.[156] Péclet discussed the need for a common source of small size and also mentioned the condition of path difference.[157] Pouillet, a recently converted "undulationist," remarked in 1830 that the sun cannot be used as a source in some interference experiments.[158] The best French exposi-

[154] Barton, "On the inflexion of light," *The Philosophical Magazine* 2 (1833): 264–65.
[155] Despretz, *Traité de physique*, 603.
[156] Babinet, *Résumé de la physique*, 165.
[157] Péclet, *Cours de physique*, 490, 492.
[158] Pouillet, *Elemens de physique expérimentale et de météorologie*, 2 vols (Paris, 1830), 2: 399–400.

tion of coherence, which was very close to Fresnel's, appeared in the 1832 edition of Péclet's book.[159]

The first German treatises which discussed the wave theory paid little attention to coherence. Baumgartner briefly referred to the conditions of frequency and direction, while the Weber brothers ignored coherence altogether.[160] The situation did not improve in the early 1830's. Although Brandes apparently began to be sympathetic with the wave theory, he said nothing about coherence.[161] Gustav Fechner's account of the wave theory was the best in Germany at the time, but it had only a few words on coherence: a brief remark (without any explanation) on the conditions of a common source and direction.[162]

British physicists were no more concerned with the concept of coherence than German. John Herschel, for instance, formulated the requirement for a "common source" in the following way:

> Two systems of interfering waves having their origins distant by an exact number of undulations may be regarded as having a common origin.[163]

He had no difficulty with this definition from a mathematical point of view, but physically it was erroneous, for it allowed a path difference of any length. Baden Powell was probably following Herschel when he admitted that interfering pencils may be emitted from "single points near each other."[164] Of all the conditions of coherence he introduced only that of direction. George Airy's famous book, a standard work on the wave theory of light in Britain, mentioned only the conditions of frequency and path difference, but without any comment.[165] The explanation of coherence was not essentially improved after the 1830's. Humphrey Lloyd and François Moigno (1804–1884), for instance, gave nothing on coherence except for the erroneous statement thath two interfering sets of waves may originate from two luminous origins.[166] Airy did not add anything on coherence in subsequent editions of his book. The best accounts of coherence during the 1850's and 1860's were those of Félix Billet (1808–1882) and Verdet who closely followed Fresnel.[167]

---

[159] Péclet, *Traité élémentaire de physique*, 2 vols, 2nd ed. (Paris, 1832), 2:385–94.

[160] Baumgartner, *Die Naturlehre*, 2:92–93.

[161] Brandes, "Interferenz," 770–87.

[162] Fechner, *Repertorium der Experimentalpyhsik*. . ., 3 vols. (Leipzig, 1832), 2:373.

[163] Herschel, "Light," 457.

[164] Powell, *A Short Elementary Treatise on Experimental and Mathematical Optics* (Oxford, 1833), 123.

[165] Airy, *On the Undulatory Theory of Optics* [1831] (London, 1866), 7, 46. This edition reproduced the first one without changes.

[166] Lloyd, *Elementary Treatise on the Wave Theory of Light*, 2nd ed. (London, 1857), 57.

[167] Billet, *Traité* 1:52, 59, 62–63, 444–5, 452. Verdet, *Leçons*, 1:53, 72–74, 101–107.

One may, of course, argue that these errors resulted from lack of care rather than of comprehension. However, only a misunderstanding of the concept of coherence can account for the erroneous conclusions in the following cases. A number of authors (e.g., Herschel, Lloyd, Bertrand, and Pouillet) identified Young's two-slit experiment with Grimaldi's and implied that the latter observed interference fringes.[168] Baumgartner claimed to have repeated Grimaldi's experiment and observed fringes, though not distinctly.[169] Carlo Matteucci (1811–1868) employed Grimaldi's experimental arrangement to demonstrate the interference of infrared rays.[170] He placed a hot ball before a screen with two small holes in it and held a heat detector behind the screen. It showed a slightly higher temperature when one of the holes was closed than when both of them were open, and Matteucci interpreted this as destructive interference. Whether his observation was accurate or not (its brief description does not allow us to judge it), it is important for us, for it caused a dispute between Arago and Matteucci about the role of a source's dimensions. Arago questioned Matteucci's result because of the large size of the source employed. It seems that he denied the possibility of using an extended source for interference observations.[171] Matteucci insisted on such a possibility, but he could not calculate whether the necessary conditions were fulfilled in his particular case.

Barton invokes the same problem of an extended source to refute Fresnel's theory of light.[172] He objects to the latter's statement that a small opening illuminated by the sun may be treated as a point source. In Barton's view, it is the points at the sun's surface that must be considered as the light sources. With this assumption he easily demonstrates that the path difference at any point of a white screen becomes indeterminate, for different choices of luminous points at the sun give different path differences. Powell was eager to defend the wave theory but he was unable to clarify the matter and simply referred Barton to Herschel's and Airy's treatises, which in fact did not treat the problem at all.[173]

Powell and Airy revealed one more flaw in their understanding of the concept of coherence while discussing Newton's Observation 6 on diffraction. Newton claimed that when the edges of two knives were very close to one another, a dark

[168] Herschel, "Light," 485; Lloyd, *Elemantary Treatise*, 56–57; Bertrand, *Lettres* 2:305; Pouillet, *Elemens de physique* 2:389.

[169] Baumgartner, "Über den optischen Interferenzversuch," *Zeitschrift für Physik* 7 (1830): 401.

[170] Matteucci, "Lettre . . . à M. Arago sur quelques phénomènes relatifs au calorique," *Bibliothèque Universelle* 50 (1832): 1–6.

[171] Matteucci, "Mémoire sur l'interférence des rayons calorifiques obscurs," *Bibliothèque Universelle* 57 (1834): 75.

[172] Barton, "On the inflection of light," 2:264, 3:176–77.

[173] Powell, "Remarks on Mr. Barton's reply, respecting the inflection of light," *The Philosophical Magazine* 3 (1833): 415.

fringe appeared in the middle of the bright space between them.[174] According to Fresnel's theory of diffraction, the central fringe in Newton's experiment had to be bright. Thus, Barton took Newton's result as a refutation of Fresnel's theory.[175] Powell supposed the cause of this discrepancy to be the large size of Newton's secondary source of light (an aperture about a quarter of an inch wide).[176] Airy verified this suggestion both experimentally and theoretically and concluded that the size of the aperture did not affect the fringes' positions.[177] However, Airy's derivation had a hidden flaw: he implicitly assumed that all points of the aperture are coherend sources, which is not true when the primary source (the sun) and the opening's size are taken into account.

All these mistakes resulted from an insufficient understanding of coherence. Until the 1890's physicists were concerned only with the location of interference fringes, and this problem usually could be solved by assuming a monochromatic point source of light. This circumstance led to a neglect of studies of the conditions of coherence, and very few physicists attempted to go beyond Fresnel in this field.

Hippolyte Fizeau (1819–1896) and Leon Foucault (1819–1868) tried to evaluate the length of Fresnel's wave trains in interference experiments with a large path difference.[178] Emmanuel Liais believed that interference fringes become narrower when the luminous source approaches the observer, and broader when it recedes.[179] It seems that he derived this erroneous result from the assumption that a light ray can interfere with another ray emitted by the same source at a later moment, which is wrong. In 1865, Verdet evaluated quantitatively the maximal size of a source for interference observations.[180] He deduced that vibrations, produced at a screen at the distance a from a source of radius R are coherent only within a circle of radius ρ if

$$\varrho \leqq \frac{a\lambda}{2R}, \tag{8.2}$$

174 Newton, *Opticks*, Book III, Obs.6, p.328.

175 Barton, "On the inflection of light," 2: 268.

176 Powell, "Remarks," 430.

177 Airy, "On the calculation of Newton's experiments on diffraction," *Trans. Cambr. Phil. Soc.* 5 (1835): 101–11.

178 Fizeau et Foucault, "Sur le phénomène des interférences entre deux rayons de lumière dans le cas de grandes différences de marche," *Comptes Rendus* 21 (1845): 1115–58.

179 Liais, "Sur les sources de lumière et les causes de non-interférence," *Mémoires de Societé des sciences naturelles et mathématiques de Cherbourg* 1 (1852): 175–80.

180 Verdet, *Lecons* 1: 101–107, also "Étude sur la constitution de la lumière non polarisée . . .," *Annales scientifiques de l'Ecole Normal Superior* 2 (1865): 295–6.

where λ is the wavelength. Verdet applied this equation to examine the conditions necessary to perform Grimaldi's two-opening experiment. A new stage in the development of the principle of interference started in 1889 and 1890 when a group of physicists (Lord Rayleigh, Albert Michelson, Charles Fabri, and others) began an intensive study of the role of the source's size and motion to fulfill new demands of spectroscopy and interferometry.

VIII.8 Young's role after 1815

There are two extreme views on Young's role in establishing the wave theory. According to many nineteenth-century historians, the wave theory was established by Young and Fresnel, while modern "revisionists" attribute this accomplishment to Fresnel alone. The two views use different definitions of "establishing" a theory: the former means creating a theory, whereas the latter emphasizes its dissemination. I prefer the "revisionist" view on the subject and agree with them that the establishment of the wave theory began in 1816, however, I will argue that Young influenced this process both directly and indirectly.

The first recipient of his direct influence was Fresnel: Young's ideas probably facilitated the rediscovery of the principle of interference and certainly helped him to develop the concept of transversal waves. When modifying Young's "screening" experiment Arago made one of his best discoveries in the field of interference, which played an important role in building confidence in the principle of interference. If readers benefited from Arago's formulations of the principle of interference and conditions of coherence, they had to thank Young. Fraunhofer based his theory of diffraction grating on Young' early papers, while works of Poisson and Herschel display the influence of the "Chromatics."

Two examples of the indirect influence seem to me the most important. First, I believe that the challenge created by Young stimulated Fresnel more than that of all Parisian "emissionists" combined. As Buchwald has noted, Fresnel was driven by a desire to make his mark in science.[181] He found a nice and important subject but the field was not free: whatever Fresnel tried, he heard that this had already been done by Young. This created an enormous pressure on Fresnel, forcing him to move ahead without stopping, improving old Young's theories and covering areas untouched by Young. He certainly wanted to demonstrate the truth of the wave theory, but no less than that the wanted to prove himself to be Young's equal.

[181] Buchwald, *Wave Theory*, 113, 116.

The following example will illustrate my point. Historians have paid little attention to the fact that the precision of Fresnel's diffraction experiments in his prize-winning memoir considerably exceeded the precision not only of the experiments of other physicists but also the rest of Fresnel's own optical observations. As shown above, Biot found the precision of Fresnel's 1816 diffraction experiments "extreme." Now, the question arises: if the struggle for the extreme precision was not Fresnel's natural inclination, and if his colleques were so satisfied with his early experiments on diffraction, why did he go into the trouble of devising a much more exact technique for his memoir presented for the contest on diffraction? As Fresnel explained, this was necessary to prove that his new theory of diffraction, based on Huygens principle, better corresponds observations than the old one. What he did not say was that the old theory was Young's theory and that was the first opportunity for Fresnel to show that he could do something better than Young. As to the exactness of measurements, Fresnel was as pragmatic as Young. He was fully satisfied with his diffraction experiments presented in his 1816 paper on diffraction, the precision of which was no better than that of Young's experiments (see VII.2). In my view, if the theory of diffraction applied in the 1816 paper were his own and not Young's, Fresnel would hardly have tried to improve it instead of spending time on other subjects of interest to him.

The other example of Young's indirect influence can be seen in the steady rise of popularity of the principle of interference, especially in Paris, after March 1816. As shown above, there was nothing magic in Fresnel's paper to change the attitude towards the principle of interference. Nonetheless, this attitude changed and quite rapidly: from a total indifference to an active interest in the dispute over the priority.[182] As shown above, this change was affected by an interest of French mathematical physicists in possible applications of this principle.

I would conjecture another possible reason, which is very difficult to document: a crazy idea sounds not so crazy when repeated a second time. I believe that both Young and Fresnel benefited from that repetition. The subject requires an additional investigation, and I will mention here only one fact. After reading the manuscript of Fresnel's first paper Arago began repeating Young's "screening" experiment (of which Fresnel had been unaware yet) to verify it. He succeeded, and this Young's experiment (not Fresnel's!) became for him (and later for others) a *decisive proof* of the principle of interference.

---

[182] See, for instance Fresnel's letters to his brother Leonor of September 25, and October 14, 1816; *Oeuvres* 2: 836–37.

[183] See, for instance, Schweigger's note in *Journal für Chemie und Physik* 19 (1817): 80n; Haüy, *Traité de physique*, 3rd ed (Paris, 1821), 2: 381–82; *Bibliothèque Universelle* 12 (1819): 174n; and J. Herschel, "Light", 449, 456.

Thus, Young's works prepared the ground for the success of Fresnel's first paper and continued to help the cause of the wave theory even after Fresnel came to dominate the field. This was recognized by Fresnel's contemporaries who always mentioned Young together with Fresnel and Arago as the creators and defenders of the new wave optics. And if they did so, I see no reason why we should do otherwise.

VIII.9 Summary

Even fifty years after Fresnel the physical aspect of the principle of interference was still insufficiently understood. Few physicists exceeded Young's level of understanding of coherence, the majority treated the principle of interference only as a mathematical concept: they concentrated on calculations of fringes' positions and did not pay any attention to their visibility. This tradition originated at the time when several French academicians accepted the principle of interference simply as a means to mathematize the phenomena of periodical colors, independent of hypotheses on the nature of light. They considered the hypotheses employed to be of secondary importance to the mathematical laws of nature obtained with their help. Consequently, the eminent academicians-"emissionists" encouraged Fresnel's work and disregarded his dissident views on the nature of light. By approving the principle of interference at an early stage, these influential scientists assisted the quite rapid incorporation of this physical concept into optics. Physicists accepted the principle of interference without a proper understanding of its physical meaning, simply because it provided a good agreement with experiment. Such a mathematical approach to the concept of interference, so beneficial to it at the early stage, became its drawback late in the nineteenth century.

Before 1827, almost all the creative work on interference had been done by Young and Fresnel; among others, the "undulationist" Fraunhofer and the "emissionist" Poisson both added to the theory, while Arago's contribution was limited to experiment. The acceptance of the principle of interference became the major condition for the subsequent establishment of the entire wave theory. Since in the 1820's the wave theory of light was identified with the theory of interference, we may say that some "emissionist" deserve credit for assisting – though unintentionally – in the establishment of the wave theory at its early stage.

Conclusions

The concept of interference was difficult to discover and proved still more difficult to grasp. It took Young three years to proceed from the principle of superposition of motions, which he borrowed from his predecessors, through the principle of superposition of waves to the principle of interference, first of sound and then of light, all of which he discovered independently. The principle's "obvious" association with waves was questioned for a quarter of a century even in mechanics and acoustics, while its physical meaning in optics continued to baffle scientists for several decades after Fresnel.

Initially, the principle of interference was almost unanimously rejected. As the comparison with Fresnel's theory shows, this could not have resulted from such alleged deficiencies of Young's theory as the "abundance of hypotheses" or the "low precision" of his experiments. Besides, none of the contemporaries criticized Young's works or attempted to deny his priority on such grounds as a poor agreement between the theory and experiment.

There were other internal causes, however, which could have had a hindering effect on Young's readers. The famous terseness of Young's writing probably discouraged many, but not all, from following his mathematics or reasoning. Those who did manage to reconstruct his equations (Mollweide) probably failed to understand the physical aspect of the concept of interference, in particular, the rare occurrence of interference phenomena and an apparent violation of the laws of photometry. However the available evidence of this comes from a later period; in the 1800s scientists struggled with a much simpler issue, namely, whether the principle of interference does not contradict Newton's theory of colors. Still others neglected the principle of interference because they failed to explain it within the emission theory and refused to accept the wave theory. While Fresnel's mathematics did not pose a problem to his readers, the other two factors did, for his explanation of the physical aspect of interference was not sufficiently clear, and the bias against the wave theory persisted from the 1800s.

As to external causes, Newton's authority remained the same in 1816 as in 1801; Brougham's strictures could have swayed only the general public and some "non-experts," but not the most reputable scientists whose approbation Young was seeking, and who were also not afraid of deviating from Newton. It appears, however, there was an external cause, which has not yet been discussed, the effect of which was different in 1801 and in 1816. I will call it the "audience factor."

Since Young's theory was mathematical, it primarily targeted the "quantitative physicists," in particular those who were interested in explaining periodical colors. Actually, Young needed only a handful of scientists who were interested in his theory, able to understand it, and possibly able to develop it further. However, in the 1800s Wollaston was the only active "quantitative physicist" in Britain, who tried to encourage Young but did not study periodical colors and never applied his principle of interference. Although French scientists had a lead over British in quantifying physics, that was not the case in physical optics before 1805. Thus, when Young's theory appeared it did not have the audience it needed.

By 1816, the situation improved in Paris. The trend towards mathematization of physics dominated the Academy of Sciences, and through the efforts of Laplace, Malus, Arago, and Biot physical optics was transformed into a mathematical science. Between 1816 and 1822, several prominent "quantitative physicists," most of whom were the "emissionists," approved the principle of interference as a means for mathematizing diffraction and chromatic polarization, and their opinion swayed other scientists. They ignored the factors which were important to the "qualitative physicists," such as the bias against the wave theory or lack of understanding of the physical aspect of the concept of interference. With no other positive factors in sight we have to conclude that it was primarily the audience factor which won the acceptance of the principle of interference.

If this were true, one may ask, why did not these scientists accord the principle of interference the same degree of acceptance several years earlier? I can offer two suggestions. As to those few (Malus, Arago, and Biot) who read Young's works and were involved in physical optics, in 1811–12 they were preoccupied with investigations of polarization, which appeared to be unconnected with Young's theory. Two years later, Young found a connection; however by that time Malus was dead, Arago apparently did not receive Young's paper, and Biot was so content with his own theory that he refused to study Young's. Young's 1814 theory of chromatic polarization was his best opportunity to attract attention of French physicists but he had too little time for that before the advent of Fresnel. Thus, when attention to a new theory is expected from a very few people acting individually much is left to chance. Even the competition factor which set Biot against reading Young is not as predictable as it appears, for after 1816 Biot's attitude towards theory of interference changed although the competition did not stop.

On the other hand, Fresnel's principle of interference received not only individual but also an institiutional approval (in 1816, 1819, and 1822). Young could not have obtained it from the Royal Society, which did not allow any discussions of papers read and never pronounced on them. The Royal Society certainly could have encouraged Young by awarding him a medal but chose not to do so even after he was recognized in Paris. Since Biot turned down Young's paper on chromatic polarization in 1815, its chances to be read to the Academy of Sciences were nil. Hence, it was unlikely for Young to receive either individual or institutional support in Paris before 1816.

The factors ignored by the "quantitative physicists" in 1816 were the only ones of importance in 1801, because there were almost no "quantitative physicists" in physical optics then, while the "qualitative physicists" viewed these factors as undermining Young's position. Hence, Young's principle of interference failed because it appeared at the wrong time and the wrong place.

In 1801, the principle of interference did not help the cause of the wave theory because the "qualitative physicists" were unable to break out of the impasse of which of the two was to be proven first. That was accomplished by the "quantitative physicists" (mostly the "emissionists"). Between 1816 and 1819 several prominent "emissionists" realized that theory of interference could explain quantitatively many phenomena unexplained in the emission theory, and the choice they faced was *either adopting theory of interference or leaving physical optics*. Thus, they approved theory of interference solely as a phenomenological theory. The association of the principle of interference with the wave theory raised the status of the latter to the level of the emission theory. Scientists began comparing the merits of the two theories, something which had never happened in the 1800s, and some were leaning towards the wave theory.

Thus, the modern wave theory actually entered optics through the back door, having been aided (albeit unintentionally) by some of its staunchest opponents. While the acceptance of the principle of interference paved the way for the wave theory in the 1820s, the initial rejection of the principle was only one of the reasons for the lack of interest in the wave theory in the 1800s. The full solution of the latter problem still awaits its investigators.

Appendix

Table 1

Wavelengths calculated from Newton's measurements

COLOR	fringe order									
	with phase change								without phase change	
	1		2		3		4		2	
Extreme			2.62	665	2.56	650			1.97	500
RED	3.60	914	2.44	620	2.32	589	2.30	584	1.83	465
Intermediate										
ORANGE	3.20	813	2.30	584					1.72	437
Intermediate										
YELLOW	2.84	721	2.17	551					1.63	414
Intermediate							2.06	523		
GREEN			2.02	513			2.02	513	1.51	384
Intermediate							1.94	494		
BLUE	0.96	244	1.87	475	1.87	475			1.40	356
Intermediate										
INDIGO			1.71	434	1.77	450			1.28	325
Intermediate					1.68	427				
VIOLET			1.49	378					1.12	284
Extreme										

Each wavelength is given in terms of 10^{-5} in and nm. Colors with no results are retained for a comparison with Table 2.

Table 2

Wavelengths calculated from Newton's measurements by Young, Biot, and Fresnel compared with the modern values

COLOR	Young		Biot		Fresnel	Modern
Extreme	2.66	676	2.54	645	645	700
RED	2.56	650			620	650
Intermediate	2.46	625	2.35	596	596	610
ORANGE	2.40	610			583	600
Intermediate	2.35	597	2.25	571	571	590
YELLOW	2.27	577			551	580
Intermediate	2.19	556	2.10	532	532	570
GREEN	2.11	536			512	520
Intermediate	2.03	516	1.94	492	492	500
BLUE	1.96	498			475	470
Intermediate	1.89	480	1.81	459	459	
INDIGO	1.85	470			449	450
Intermediate	1.81	460	1.73	439	439	
VIOLET	1.74	442			423	410
Extreme	1.67	424	1.60	406		400

Young's and Biot's results are given in terms of 10^{-5} in and in nm, while Fresnel's and modern data are in nm only.

Table 3

Diameter of water drops in a cloud calculated on the basis of Young's formula from the measurements of Newton and Jordan

Observation	fringe order m	angular radius Θ	$\tan\Theta$	$d \times 10^{-5}$ in
Newton	1*	3.63	0.063	41
observ. 1	2*	5.13	0.090	57
	3	6.17	0.108	71
	1*	2.75	0.048	50
observ. 2	2*	4.67	0.082	59
	3	6.0	0.105	69
observ. 3	1	1.5	0.026	88
	2	2.75	0.048	95
Jordan	1	1.0	0.017	150
observ. 1	2	1.6	0.029	176
observ. 2	1*	2.8	0.049	52
	2*	5.0	0.087	59
Young (Lectures)				$d > 50$

* Observations which satisfy Young's condition (the distance between the adjacent rings is about 2°–3°).

Table 4

Wavelengths calculated from observations on diffraction presented in Young's "Experiments and Calculations."

Experiment	Fringe	wavelengths calculated by Young	*wavelengths recalculated*		
			without phase change		with phase change
		$\times 10^{-5}$ in	$\times 10^{-5}$ in	nm	nm
Newton		2.44	2.44	620	310
		3.10	3.10	787	394
	1st min	3.64	3.65	927	464
Obs. 9		3.34	3.34	848	424
		3.32	3.33	846	423
		3.32	3.32	843	422
Obs. 3	2nd max	3.02	3.02	767	511
	2nd max	3.46	3.47	881	587
	3rd max	2.60	2.61	663	530
	3rd max	2.86	2.86	726	581
Young Obs. 3	2nd min	2.98	2.98	757	568
Obs. 4	1st min	3.92	3.92	996	498
	2nd min	2.76	2.76	701	526
	3rd min	2.58	2.58	655	546

Table 5

Wavelengths calculated from the measurements of external fringes

Experiment	original results			corrected or reinterpreted results		
	λ	s	s/λ %	λ	s	s/λ %
Newton, Obs. 9	812	104	12.8	406	52	12.8
Newton, Obs. 3	758	92	12.1	552	38	6.8
Young, Obs. 3 & 4	777	152	19.6	535	30	5.6
Young, Obs. 3 & 4*				528	25	4.7
Average of all observations used by Young	783	27	3.4**			
Fresnel (1815) (Grimaldi's method)				488	33	6.8
Fresnel (1815) (Newton's method)				518	39	7.5
Fresnel (1816) (lens-micrometer)				517	36	6.9
Fresnel (1816) (red light)				604	28	4.7

* In Young's Observation 4 the wavelengths are calculated for every position of the fringes, whereas Young himself calculated them for the average of several positions of the same fringe.

** Young would have obtained this precision by treating Newton's and his own results as received in three independent experiments.

Bibliography

List of abbreviations

Ann. de Chim.	– Annales de Chimie; Annales de Chimie et Physique
Ann. der Phys.	– Annalen der Physik
Hist. Acad. Berlin	– Histoire de l'Académie royale des sciences et belle-lettres. Berlin.
Hist. Acad. Paris	– Histoire de l'Académie royale des sciences avec les mémoires de mathématique et physique. Paris
Mém. Acad.	– Mémoires de l'Académie des sciences. Paris
Mém. Sav. Etrang.	– Mémoires de mathématique et de physique, presentés a l'Académie royal des sciences, par divers savans. Paris
Nich. Jour.	– A Journal of Natural Philosophy, Chemistry, and the Arts
Phil. Trans.	– Philosophical Transactions of the Royal Society of London
Trans. Cambr. Phil. Soc.	– Transactions of Cambridge Philosophical Society

Académie des Sciences, Paris. Procès-verbaux des séances tenue depuis la foundation de l'Institut jusqu'au mois d'août 1835. 10 vols. Hendaye: Imprimerie de l'Observatoire d'Abbadia, 1910–22.

Alembert, Jean Le Rond d' "Fondamental." *Encyclopédie ou Dictonnaire raisonné des sciences, des arts et des métiers.* 17 vols. Paris, 1751–65. See v. 7, 54–63.

Ampère, André-Marie. *Correspondance du grand Ampère.* Ed. by Louis de Launey. 3 vols. Paris: Gauthier-Villars, 1936.

–. "Démonstration d'un théorème d'où l'on peut déduire toutes les lois de la réfraction ordinaire et extraordinaire [1815]." *Mém. Acad.* 1813–15: 235–48.

–. "Mémoire sur la détermination de la surface des ondes lumineux dans un milieu dont l'élasticité est différent suivant les trois directions principales.." *Annales de Chimie* 39 (1828): 113–45.

–. "Mémoire sur la théorie mathématique des phénomènes électrodynamiques uniquement déduite de l'expérience, dans lequel se trouvent reunis les mémoirs que M. Ampère a communiqués à l'Académie royale des Sciences . . ." *Mém. Acad.* 6 (1823): 175–388. Published in 1826.

Airy, George Biddell. *On the Undulatory Theory of Optics.* London: Macmillan, 1866. First published in his *Mathematical Tracts on the Lunar and Planetary Theories, the Figure of the Earth, Precession and Nutation, the Calculus of Variations, and the Undulatory Theory of Optics.* Cambridge: Deighton, 1831.

–. "On a Remarkable Modification of Newton's Rings [1831]." *Trans. Cambr. Phil. Soc.* 4 (1833): 279–88; "On the phenomena of Newton's rings when formed between two transparent substances of different refractive powers." Ibid, 409–21.

–. "Remarks on Mr. Potter's experiment on interference." *The Philosophical Magazine* 2 (1833): 161–67.

–. "On the calculation of Newton's experiments on diffraction." *Trans. Cambr. Phil. Soc.* 5 (1835): 101–11.

–. "On the intensity of light in the neighbourhood of a caustic." *Trans. Cambr. Phil. Soc.* 6 (1838): 379–402, 8 (1840): 595–600.

–. *Autobiography of Sir George Biddell Airy*, etc." Cambridge: Cambridge University Press, 1896.

The Annals of Philosophy. Vols. 1, and 3 (1801 to 1804).

The Annual Review. Vols. 1, 3, and 7 (1802 to 1808).

Arago, Dominique François. *Œuvres complètes de François Arago.* Edited by J. A. Barral. 17 vols. (Paris: Gide, 1854–62).

–. "Mémoire sur la vitesse de la lumière [1810]." First published *Annales de Chimie* 37 (1853): 180–96.

–. "Mémoire sur une modification remarquable qu' éprouvent les rayons lumineux dans leur passage à travers certain corps diaphanes d'optique." *Mém. Acad.* 1811: 93–134.

–. "Note sur un phénomène remarkable qui s'observe dans la diffraction de la lumière." *Annales de Chimie* 1 (1816): 199–202.

–. "Expérience sur la diffraction." *Bulletin de Société Philomatique* 1816: 56–57.

–. "Remarques sur l'influence mutuelle de deux faisceaux lumineux qui se croisent sous un très-petit angle." *Annales de Chimie* 1 (1816): 332–34.

–. "Rapport fait à la Première Classe de l'Institut, le 25 mars 1816, sur un mémoire relatif aux phénomènes de la diffraction de la lumière par M. Fresnel." *Œuvres de Fresnel* 1: 79–87.

–. "Sur les couleurs des lames minces." *Mémoires d'Arcueil* 3 (1817): 323–370. Reprint. New York: Johnson Reprint Corp., 1967.

–, and A. Fresnel. "Mémoire sur l'action que les rayons de lumière polarisés exercent les uns sur les autres." *Annales de Chimie* 10 (1819): 288–306.

–. "Rapport fait par M. Arago à l'Académie des Sciences au nom de la commission qui avait eté chargée d'examiner les mémoires envoyés au concours pour le prix de la diffraction." *Annales de Chimie* 11 (1819): 288–306.

–. "Examen des remarques de M. Biot." *Annales de Chimie* 17 (1821): 258–73.

–. "Rapport fait à l'Académie des Sciences, le lundi 4 juin 1821, sur un mémoire de M. Fresnel relatif aux couleurs des lames cristallisées douées de la double refraction." *Annales de Chimie* 17 (1821): 80–102.

–. "Notes sur les phénomènes de la polarisation de la lumière [1812]." *Œuvres d'Arago* 10: 75–84; "Mémoire sur plusieurs nouveaux phénomènes d'optique [1812]." Ibid, 85–97; and "Quatrième mémoire sur plusieurs nouveaux phénomènes d'optique [1812]." Ibid, 98–122.

–. "Histoire de ma jeunesse." *Œuvres d'Arago* 1: 1–102.

–. "Fresnel [1830]." *Œuvres d'Arago* 1: 107–86.

–. "Thomas Young [1832]." *Œuvres d'Arago* 1: 241–94.

–, and A. Petit. "Sur les puissances réfractives et dispersives de certain liquides et des vapeurs qu'ils forment." *Annales de Chimie* 1 (1816): 1–9.

[–]. "Question de physique, proposée par l'Acad. roy. des Sc. de Petersbourg," *Bulletin des Sciences Mathématiques* 7 (1827): 211–5.

Babinet, Jacques, and Charles Bailly. *Résumé complet de la physique des corps imponderable*. 2 vols in 1. Paris, 1825.

Babinet. "Sur les couleurs des reseaux [1827]." *Annales de Chimie* 40 (1829): 166–71.

–. "Mémoire sur les couleurs des doubles surfaces à distance." *Comptes Rendus* 7 (1838): 694–98.

Barton, John. "On the inflexion of light." *The Philosophical Magazine* 2 (1833): 263–69.

–. "On the inflexion of light, in reply to Professor Powell." *The Philosophical Magazine* 3 (1833): 172–78.

Baumgartner, Andreas. *Die Naturlehre nach ihrem gegenwärtigen Zustande mit Rücksicht auf mathematische Begründung*. 3 vols. Vienna: Heubner, 1824.

—. "Die circulare Polarisation des Lichtes nach Fresnel's Arbeiten." *Zeitschrift für Physik* 2 (1827): 1–20.

—. "Ein Beitrag zur Theorie der Beugung des Lichtes." *Zeitschrift für Physik* 3 (1827): 443–51.

—. "Über den optischen Interferenzversuch." *Zeitschrift für Physik* 7 (1830): 399–405.

Beccaria, Giacomo Battista. "Observation sur la double réfraction du crystal de roche." *Journal de Physique* 2 (1772): 504–10.

Bérard, Jacques-Étienne. "Mémoire sur les propriété des différentes espèces de rayons qu'on peut separer au moyen du prisme de la lumière solaire [1812]." *Mémoires d'Arcueil* 3 (1817): 1–47.

Bernoulli, Daniel. "Réflexions et éclaircissment sur les nouvelles vibrations des cordes exposées dans les mémoires de l'Académie de 1747 & 1748." *Hist. Acad. Berlin* 9 (1753): 147–72.

—. "Sur le mélange de plusieures espèces de vibrations simples isochrones, qui peuvent coexister dans un même système de corps." *Hist. Acad. Berlin* 9 (1753): 173–95.

—. "Recherches physiques, mécaniques et analytiques, sur le son & sur les tons des tuyaux d'orgues différemment construits." *Hist. Acad. Berlin* 1762: H 170–81, M 431–85.

Bertrand, Alexandre Jacques. *Lettres sur la physique.* 2 vols. Paris: Bossange, 1824–25.

Berzelius, Jons Jacob. *Jac. Berzelius Bref.* Ed. H. G. Soderbaum. 6 vols. & 3 suppls. Uppsala, 1912–61.

Beudant, François-Sulpice. *Essai d'un course élémentaire et général des sciences physiques.* 3rd ed. Paris: chez Verdiere, 1824.

Billet, Felix. *Traité d'optique physique.* 2 vols. Paris: Bachelier, 1858–59.

Biot, Jean-Baptiste. "Mémoire sur un nouveau genre d'oscillation que les molécules de la lumière éprouvent en traversant certains cristaux." *Mém. Acad.* 1812: 1–371.

—. "Mémoires sur les rotations que certaines substances impriment aux axes de polarisation des rayons lumineux." *Mém. Acad.* 2 (1817): 41–136. Published in 1819.

—. "Sur l'aimantation imprimée aux metaux par l'électricité en mouvement." *Journal des Savants* 1821: 281–90.

—. "Remarques sur un rapport de MM. Arago et Ampère." *Annales de Chimie* 17 (1821): 225–58.

—, and Claude Pouillet. "Sur la diffraction de la lumière." *Bulletin de Société Philomatique* 1816: 60–61.

—. *Mélanges scientifiques et litteraires.* 3 vols. Paris: M. Levy, 1858.

–. "Additions à l'optique." in *Physique mécanique, par E. G. Fischer . . . traduit de l'allemand; avec des notes, et un appendice sur les anneaux colorés, sur la double réfraction et sur la polarisation de la lumière, par M. Biot.* 3rd ed., rev. and enl. Paris: Mme. Vve. Courcier, 1819.

–. *Traité de physique expérimentale et mathématique.* 4 vols. Paris: Deterville, 1816.

–. *Précis élémentaire de physique expérimentale.* 2 vols. Paris: Deterville, 1817. See also the 3rd ed. Paris: Deterville, 1824.

Born, Max, and Emil Wolf. *Principles of Optics.* New York: Pergamon Press, 1959.

Boyer, Carl Benjamin. *The Rainbow from Myth to Mathematics.* New York: T. Yoseloff, 1959.

Brandes, Heinrich Wilhelm. "Über die farbigen Bogen, welche man zuweilen an der innern Seite des Regenbogens bemerkt." *Annalen der Physik* 19 (1805): 464–75.

–. "Auszug . . . über Parrot's Theorie der Beugung des Lichtes." *Annalen der Physik* 47 (1814): 212–18.

–. "Venturi's Theorie des farbigen Bogens, welcher sich oft an der innern Seite des Regenbogens zeigt, dargestellt mit einigen Anmerkungen." *Annalen der Physik* 52 (1816): 385–97.

–. "Inflexion des Lichtes," "Interferenz." In *Johann Gehler's Physikalisches Wörterbuch, neu bearbeitet von Brandes, Gmelin, Horner, Muncke u. Pfaff.* 11 vols. Leipzig: E. Schwickert, 1825–45. See v. 5 (1830): 681–742, 770–787.

B[rewster], D[avid]. "Literary and scientific notices. Domestic and foreign." *The Edinburgh Magazine* 20 (1802): 63

–. "Progress of manufactures, science, and fine arts." *The Edinburgh Magazine* 21 (1801): 206.

[–]. "Royal Society of Edinburgh." *Ouarterly Journal of Science* 2 (1817): 207.

[–]. "Account of M. Fresnel's discoveries respecting the inflexion of light." *The Edinburgh Philosophical Journal* 2 (1820): 150–53.

Brewster. "On the new properties of light exhibited in the optical phenomena of mother-of-pearl, and other bodies to which the superficial structure of that substance can be communicated." *Phil. Trans.* 1814: 397–418.

–. "On a new species of coloured fringes produced by the reflection of light between two plates of glass of equal thickness." *Transactions of the Royal Society of Edinburgh* 7 (1815): 435–44.

–. "On the communication of the structure of doubly-refracting crystals to glass, fluor spar, and other substances by mechanical compression and dilatation." *Phil. Trans.* 1816: 156–78.

–. "On the laws of polarization and double refraction in regularly crystallized bodies." *Phil. Trans.* 1818: 199–272.

–. "On the undulations excited in the retina by the action of luminous points and lines." *The Philosophical Magazine* 1 (1832): 169–74.

The British Critic. Vols. 19, 22, 23, 25 and 30 (1803 to 1807).

Brougham, Henry. "Experiments and observations on the inflection, reflection, and colours of light." *Phil. Trans* 1796: 227–77.

–. "Further experiments and observations on the affections and properties of light." *Phil. Trans.*, 1797: 352–85.

[–]. "Bakerian lecture on light and colours." *The Edinburgh Review* 1 (1803): 450–56.

[–]. "Young on colours not hitherto described." *The Edinburgh Review* 1 (1803): 457–60.

[–]. "Dr. Young's Bakerian lecture." *The Edinburgh Review* 5 (1804): 97–103.

[–]. "Wood's *Optics*." *The Edinburgh Review* 1 (1802): 158–63.

[–]. "Wollaston on prismatic reflection." *The Edinburgh Review* 2 (1803): 97–98.

[–]. "Wollaston on Iceland crystal." *The Edinburgh Review* 2 (1803): 99.

[–]. "Memoirs of the Philosophical Society of Manchester." *The Edinburgh Review* 2 (1803): 192–96.

[–]. "Count Rumford on the nature of heat." *The Edinburgh Review* 4 (1804): 399–414.

[–]. "Venturi, *Sopra i colori*." *The Edinburgh Review* 6 (1805): 20–43.

[–]. "Leslie's *Inquiry into the Nature of Heat.*" *The Edinburgh Review* 7 (1805): 63–91.

Buchwald, Jed Z. "Experimental investigation of double refraction from Huygens to Malus." *Archive for History of Exact Sciences* 21 (1980): 311–73.

–. *The Rise of the Wave Theory of Light: Optical Theory and Experiment in the Early Nineteenth Century.* Chicago: University of Chicago Press, 1989.

Buffon, Leclerc. *Histoire naturelle générale et particulière* 125 vols. in 61 (Paris, 1799–1808). See v. 13, 272–83.

Cantor, Geoffrey. "Thomas Young's lectures at the Royal Institution." *Notes and Records of the Royal Society of London* 25 (1970): 87–112.

–. "The historiography of 'Georgian' optics.". *History of Science* 16 (1978): 1–21.

–. "The changing role of Young's ether." *The British Journal for the History of Science* 5 (1970): 44–62.

–. *Optics after Newton: Theories of light in Britain and Ireland, 1704–1840.* Manchester University Press, 1983.

Cauchy, Augustin Louis. "Note sur l'optique." *Comptes Rendus* 2 (1836): 343–49.

Chaulnes, Michel Ferdinand, Duc de. "Observations sur quelques cas de la quatrième partie du deuxième livre l'Optique de M. Newton." *Hist. Acad. Paris* 1755: 136–44.

Chladni, Ernst. *Die Akustik.* Leipzig: Breitkopf & Härtel, 1802. New ed. Leipzig, 1830.

Clive, John. *Scotch Reviewers: The Edinburgh Review, 1802–1815.* London: Faber, 1957.

Coddington, Henry. *An Elementary Treatise on Optics.* Cambridge: Deighton, 1823.

Cohen, I. Bernard. "The first explanation of interference." *American Journal of Physics* 8 (1940): 99–106.

Crane, William. "Observations upon the different hypotheses that have been proposed respecting the nature of light." *The Philosophical Magazine* 46 (1815): 195–203.

The Critical Review. Vols. 34, 35, and 38 (1802–1803); 3, 6 and 12 (1804 to 1807).

Crosland, Maurice. *The Society of Arcueil: A View of French Science at the Time of Napoleon I.* Cambridge: Harvard University Press, 1967.

Dalzel, Andrew. *History of the University of Edinburgh from Its First Foundation with a Memoir of the Author by C. Innes.* Edinburgh: Edmonston & Douglas, 1862.

Davy, Humphrey. *Collected Works of Sir Humphrey Davy.* 9 vols. London: Smith, 1839. Reprint. New York: Johnson Reprint Corp., 1972.

Delambre, Jean-Baptiste. "Analyse des travaux de l'Académie royale des sciences pendant l'année 1816. Partie mathématique." *Mém. Acad.* 1 (1816): HI-CXII.

Delisle, Joseph-Nicolas. "Expériences sur la lumière et les couleurs." In his *Mémoires pour servir à l'histoire et au progres de l'astronomie, de la géographie et de physique,* 205–66. St. Petersbourg, 1738.

Despretz, Cesar Mansuete. *Traité élémentaire de physique.* Paris: Mequignon-Marvis, 1825.

Drude, Paul Karl. *Lehrbuch der Optik.* Leipzig: S. Hirzel, 1900.

Dutour, Étienne-François. "Recherches sur les phénomènes des anneaux colorées." *Mém. Sav. Étrang.* 4 (1763): 285–312.

–. "De la diffraction de la lumière." *Mém. Sav. Étrang.* 5 (1768): 635–77.

The Edinburgh Magazine. Vols. 20–22 (1802–1803).

The European Magazine 47 (1805): 130–31.

Englefield, Henry. "Experiments on the separation of light and heat by refraction. In a letter from Sir H. C. Englefield . . . to Thomas Young . . ." *Journals of the Royal Institution* 1 (1802): 202; reprinted in *Nicholson's Journal* 3 (1802): 125–30.

Euler, Leonard. *Leonhardi Euleri Opera Omnia.* 72 vols. to date. Leipzig, et al. 1911–.

–. "Sur la vibration des corps." *Hist. Acad. Berlin* 4 (1747–48): 69–85.

–. "Sur la propagation du son." *Hist. Acad. Berlin* 15 (1759): 185–209. "Supplément aux recherches sur la propagation du son," ibid. 210–40; and "Continuation des recherches sur la propagation du son," ibid, 241–64.

–. "Essai d'une explication physique des couleurs engendrées sur des surfaces extremement minces." *Hist. Acad. Berlin* 8 (1752): 262–82

Fechner, Gustav Theodor. *Repertorium der Experimentalphysik, enthaltend eine vollständige Zusammenstellung der neuen Fortschritte dieser Wissenschaft.* 3 vols. Leipzig: L. Voss, 1832.

Fizeau, Hyppolite & Leon Foucault. "Sur le phénomène des interférence entre deux rayons de lumière dans le cas de grandes différences de marche." *Comptes Rendus* 21 (1845): 1155–58.

–. "Mémoire sur le phénomène des interférence entre deux rayons de lumière dans le cas de grandes différences de marche, et sur la polarization chromatique produite par les lames épaisses cristallisées." *Annales de Chimie* 30 (1850): 146–59.

Flaugergues, Honore. "Sur la diffraction de la lumière." *Journal de Physique* 74 (1812): 125–29; 75 (1813): 16–29; 76: 142–154.

–. "Supplément a différent mémoires sur la diffraction de la lumière, publies dans le *Journal de Physique.*" *Journal de Physique* 89 (September 1819): 161–86.

Fox, Robert. "The rise and fall of Laplacian physics." *Historical Studies in the Physical Sciences* 4 (1974): 89–136.

Françon, Maurice. *Optical Interferometry.* Transl. from the French by I. Wilmanns. New York: Academic Press, 1966.

Frankel, Eugene. "Jean-Baptiste Biot: The career of a physicist in nineteenth-century France." Ph. D. diss., Prinston University, 1972.

–. "J. B. Biot and the mathematization of experimental physics in Napoleonic France." *Historical Studies in the Physical Sciences* 8 (1977): 33–72.

–. "Corpuscular optics and the wave theory of light: The science and politics of a revolution in physics." *Social Studies of Science.* 6 (1976): 141–84.

Fraunhofer, Joseph. "Neue Modifikation des Lichtes durch gegenseitige Einwirkung und Beugung der Strahlen, und Gesetze desselben." *Denkschriften der Königl. Baierischen Akad. der Wissenschaft* 1821–22: 1–76. French translation appeared in *Astronomische Abhandlungen (Altona)* 2 (1823): 46–112.

–. "Kurzer Bericht von den Resultaten neurer Versuche über die Gesetze des Lichtes und die Theorie derselben." *Annalen der Physik* 74 (1823): 337–78.

Fresnel, Augustin Jean. *Œuvres complètes d'Augustin Fresnel.* Eds. Henri de Senarmont, Emile Verdet, and Léonor Fresnel. 3 vols. Paris: Imprimerie impériale, 1866–70. Reprint. New York: Johnson Reprint Corp., 1965.

–. "Mémoire sur la diffraction de la lumière, où l'on examine particulièrement le phénomène des franges colorées que présentent les ombres des corps éclairés par un point lumineux [October 15, 1815]." *Œuvres* 1:9–33.

–. "Supplément au mémoire sur la diffraction [November 10, 1815]." *Œuvres* 1:41–60.

–. "Mémoire sur la diffraction de la lumière [1816]." *Annales de Chimie* 1 (March 1816): 239–81.

–. "Mémoire sur la diffraction de la lumière (extrait)." *Annales de Chimie* 11 (1819): 246–96, 337–78.

–. "Mémoire sur la diffraction de la lumière, couronné par l'Académie des Sciences." *Mém. Acad.* 5 (1821–22). Published in 1826.

–. "Supplément au deuxième mémoire sur la diffraction de la lumière [July 15, 1816]." *Œuvres* 1: 129–70.

–. "Mémoire sur l'influence de la polarisation dans l'action que les rayons lumineux exercent les uns sur les autres [two versions dated August 30, and October 6, 1816]." *Œuvres* 1: 385–440.

–. "Note sur les phénomènes de la diffraction dans la lumière blanche [n. d.]." *Œuvres* 1: 193–94.

–. "Extrait d'une lettre de Fresnel à Arago, sur l'influence de la chaleur dans les couleurs développées par la polarisation." *Annales de Chimie* 4 (1817): 298–300.

–. "Lettre à François Arago sur l'influence du mouvement terrestre dans quelques phénomènes d'optique." *Annales de Chimie* 9 (1818): 57–66, 286.

–. "Note sur l'application du principe d'Huygens et de la théorie des interférences aux phénomènes de la réflexion et de la réfraction [1819]." *Œuvres* 1: 201–19.

–. "Note sur le calcul des teintes que la polarisation développe dans les lames cristallisées." *Annales de Chimie* 17 (1821): 101–12, 167–96, 312–16.

–. "Mémoire sur les modifications que la réflexion imprime a la lumière polarisée [November 10, 1817]." *Œuvres* 1: 441–86.

–. "Mémoire sur les couleurs developpées dans les fluides homogènes par la lumière polarisée [March 30, 1818]." First published *Annales de Chimie* 17 (1846): 172–99.

–. "Mémoire sur la double réfraction." *Mém. Acad.* 7 (1824)ß 45–176. Published in 1827.
–. "Résumé d'un mémoire sur la réflexion de la lumière." *Annales de Chimie* 15 (1809): 379–86.
–. "De la lumière." In *Supplément* to T. Thomson. *Système de Chimie.* 4 vols. & suppl. 5th ed. Translated by J. Riffault. Paris, 1822.
–. "Réponse de M. A. Fresnel à la lettre de M. Poisson . . . 'Sur le vibrations d'une onde lumineuse.'" *Annales de Chimie* 23 (1823): 32–49, 113–23.
–. "Quelques observations sur les principales objections de Newton contre le système des vibrations lumineuses et sur les difficultes que presente son hypothèse des accès." *Bibliothèque Universelle* 22 (1823): 73–99
–. "De la lumière. Par Mr. A. Fesnel . . . Mémoire inseré dans le volume supplémentaire de la traduction française de la *Chimie* de Thomson (Extrait)." *Bibliothèque Universelle* 22 (1823): 3–17.
–. "Über das Licht, von Hrn. Fresnel." *Annalen der Physik* 3 (1825): 89–128, 303–28; 5: 223–56; 12 (1828): 197–249, 366–99.
–. "Elementary view of the undulatory theory of light by M. Fresnel, from the *Supplement* to the translation of Thomson's *Chemistry*." *The Quarterly Journal of Science* 1 (1827): 127–41, 441–54; 2: 113–35; 3 (1828): 198–215; 4: 168–91, 389–407; and 5 (1829): 159–65.
Fuss, Paul. "Sur le prix de physique proposé par l'Acad. Imp. de Sci. de St. Petersbourg." *Bulletin des Sciences Mathématiques.* 9 (1828): 96–108.
Gough, John. "The theory of compound sounds." *Memoirs of the Literary and Philosophical Society of Manchester* 5, pt. 2 (1802): 653–65; also *Nicholson's Journal* 4 (1803): 152–59.
–. "Reply to Dr. Young's letter on the theory of compound sounds." *Nicholson's Journal* 3 (1802): 39–41.
–. "On the nature of the grave harmonics." *Nicholson's Journal* 4 (1803): 1–4.
–. "On the nature of musical sounds. In reply to Dr. Young." *Nicholson's Journal* 4 (1803): 139–42.
–. "Experiments and remarks on the augmentation of sound." *Nicholson's Journal* 10 (1805): 65–69.
–. "A mathematical theory of the speaking trumpet." *Nicholson's Journal* 10 (1805): 160–65.
Grattan-Guinness, Ivor. "Essay reviews. Recent researches in French mathematical physics in the early 19th century." *Annals of Science* 38 (1981): 663–90.

–. *Convolutions in French Mathematics, 1800–1840* Basel. Birkhäuser, 1990.

Gravesande, Willem Jacob van s.' *Mathematical Elements of Natural Philosophy.* Transl. into English by J.Desaguliers. 2 vols. 5th ed. London: J.Senex, 1737.

Hargreave, David. "Thomas Young's theory of color vision: its roots, development, and acceptance by the British scientific community." Ph. D. diss., University of Wisconsin, 1973.

Hassenfratz, Jean Henri. "Sur la cause qui augmente l'intensité du son dans les porte-voix." *Annales de Chimie* 50 (1804): 297–311.

Haüy, René-Just. *Traité élémentaire de physique.* 2 vols. Paris: Imprim. Delance & Lesueur, 1803. Also see 3rd ed. rev. & cons. augm. Paris: Bachelier, 1821.

Heilbron, John. *Electricity in the 17th and 18th Centuries.* Berkeley: University of California Press, 1979.

Helmholtz, Hermann. "Über Combinationstöne." *Annalen der Physik* 99 (1856): 497–540.

Herschel, John Frederick William. "Light [1827]." In *Encyclopaedia Metropolitana.* 26 vols. London, 1817–1845. Vol.4, 341–586.

–. *Traité de la lumière.* Trans. into French by P. Verhulst & A. Quetelet. 2 vols. Paris, 1829 (v.1); Paris, 1833 (v.2).

–. *Vom Licht.* Translated into the German by J. E. Schmidt. Stuttgart: J.G.Cotta'schen Buchhandlung, 1831.

Herschel, William. "Experiments for investigating the cause of the coloured concentric rings, discovered by Sir Isaac Newton, between two object-glasses laid upon one another." *Phil. Trans.* 1807: 180–233; "Continuation of experiments ..." Ibid. 1809: 259–302; and "Supplement ..." Ibid. 1810: 149–77.

Hooke, Robert. *Micrographia: or, Some Physiological Descriptions of Minute Bodies Made by Magnifying Glasses, with Observations and Inquiries Thereupon.* London, 1665. Reprint. New York: Dover Publications, 1961.

–. *The Posthumous Works of Robert Hooke.* London, 1704. Reprint. New York: Johnson Reprint Corp., 1969.

Humboldt, Alexandre, and A. Bonpland. *Voyage aux régions équinoxiales du nouveau continent, fait en 1799, 1800, 1801, 1802, 1803 et 1804 par Al. de Humboldt et A. Bonpland, redigé par Alexandre de Humboldt.* 12 vols. Paris: Librarie grecque-latine-allemande, 1815–26.

Huygens, Christiaan. *Treatise on Light.* Translated by Sylvanus P. Thompson. University of Chicago Press, [1912].

The Imperial Review. Vols. 2, and 3 (1804).

Jenkins, F., and H. White. *Fundamentals of Optics*. 3rd ed. New York: McGraw-Hill, 1957.

Jones, Henry Bence. *The Royal Institution, its Founder and Its First Professors*. London: Longman et al., 1871. Reprint. New York: Arno Press, 1975.

[Jordan, Gibbes Walker]. *The Observations of Newton Concerning the Inflexion of Light Accompanied by Other Observations Differing from His; and Appearing to Lead to a Change of His Theory of Light and Colours*. London: T. Cadell & W. Davies, 1799.

[–]. *An Account of the Irides or Coronae Which Appear Around, and Contiguous to the Bodies of the Sun, Moon, and Other Luminous Objects*. London: T. Cadell & W. Davies, 1799.

[–]. *New Observations Concerning the Colours of Thin Transparent Bodies* . . . London: T. Cadell & W. Davies, 1800.

Journals of the Royal Institution Vols. 1–2 (1802–1803).

Knox, John. "On some phenomena of colours exhibited by thin plates." *Phil. Trans.* 1815: 161–81.

Lagrange, Joseph-Louis. *Œuvres*. 14 vols. Paris, 1867–92.

Lambert, Johann. "Sur quelques instruments acoustiques." *Hist. Acad. Berlin* 19 (1763): 87–124.

Lamé, Gabriel. *Cours de physique de l'École Polytechnique*. 2 vols. Paris: Bachelier, 1836–37.

Laplace, Pierre-Simon. "Sur le mouvement de la lumière dans milieux diaphanes." *Mémoires d'Arcueil* 2 (1809): 111–42.

– . *Traité de mécanique céleste*. 6 vols. Paris: Duprat, 1798–1823. Vol. 4 (1805).

La Provostay, Frederic Hervé de, and Paul Desains. "Mémoire sur les anneaux colorés." *Annales de Chimie* 28 (1849): 423–39.

Latchford, Kenneth. "Thomas Young and the evolution of the interference principle." Ph. D. diss., London Imperial College, 1975.

Liais, Emmanuel. "Sur les sources de lumière et les causes de non-interférence." *Mémoires de Societé des sciences naturelles et mathématiques de Cherbourg* 1 (1852): 175–80.

Libri, Guglielmo. "Les sciences en France." *Revue de deux Mondes* 21 (1840): 789–819.

Lipson, S. G., and H. Lipson. *Optical Physics*, 2nd ed. Cambridge University Press, 1981.

Lloyd, Humphrey. "Report on the progress and present state of physical optics." *British Association Reports* 1834: 295–413.

– . *Elementary Treatise on the Wave Theory of Light*. 2nd ed., with adds. London: Longman et al., 1857.

Lüdicke, August. "Versuche über die Mischung prismatischer Farben." *Annalen der Physik* 34 (1810): 1–27, 229–39, 362–89.

Mach, Ernst. *The Principles of Physical Optics; an Historical and Philosophical Treatment.* Trans. by J.S. Anderson & A.F. Young. London: Methuen, 1926. Reprint. New York: Dover Publications, 1953.

Mairan, Jean-Dortous. "Recherches physico-mathématique sur la réflexion des corps. Troisième partie. De la diffraction." *Hist. Acad. Paris* 1738: 1–64.

–. "Discours sur la propagation du son dans le différens tons qui le modifient." *Hist. Acad. Paris* 1737: H 133–43, M 1–87.

Malus, Étienne-Louis. "Sur une propriété de la lumière réflechie par les corps diaphanes." *Bulletin de Societé Philomatique* 1809: 266–69.

–. "Théorie de la double réfraction." *Mém. Sav. Étrang.* 2 (1811): 303–508.

Maraldi, Giacomo Filippo. "Diverses expériences d'optique." *Hist. Acad. Paris* 1723: 157–200.

Matteucci, Carlo. "Lettre de M. Matteucci à M. Arago sur quelques phénomènes relatif au calorique." *Bibliothèque Universelle* 50 (1832): 1–6; and "Mémoire sur l'interférence des rayons calorifiques obscurs." Ibid. 57 (1834): 74–80.

Mayer, Johann Tobias. "Phaenomenorum ab inflexione luminis pendentium ex propriis observationibus et experimentis recensio et comparato." *Commentationes recentiores Societatis Goettingae* 4 (1820): 49–80.

Mayer, J. T. "Über die von der Inflexion und Deflexion des Lichtes abhandlungen Erscheinungen." *Journal für Chemie und Physik* 25 (1819): 231–37.

Mazeas, Guillaume. "Observations sur les couleurs engendrées par le frottement des surfaces planes et transparents." *Hist. Acad. Berlin* 8 (1752): 248–61.

The Monthly Magazine. Vols. 13, 14, 15, 18 and 20 (1802 to 1805).

The Monthly Review. Vols. 33, 35, 39, 40, 45 and 46 (1800 to 1805).

Mollweide, Karl. *Prüfung der Farbenlehre des Hrn. von Gothe und Vertheidigung des Newton'schen Systems gegen dieselbe.* Halle, 1810.

Morse, Edgar W. "Natural Philosophy, hypotheses and impiety: Sir David Brewster confronts the undulatory theory of light." Ph. D. diss., University of California, Berkeley, 1972.

–. "Young, Thomas." *Dictionary of Scientific Biography.* Editor in Chief C.C. Gillispie. Vol. 14, 562–72. New York: Charles Scribner's Sons, 1976.

Newton, Isaac. *The Mathematical Principles of Natural Philosophy.* Translated into English by Andrew Motte in 1729, revised by Florian Cajori. Berkeley, Cal.: University of California Press, 1946.

–. *Opticks; or A Treatise of the Reflections, Inflections and Colours of Light.* Based on the 4th ed. London, 1730. New York: Dover Publications, 1952.

–. *Isaac Newton's Papers and Letters on Natural Philosophy.* Edited by I. B. Cohen. Cambridge: Harvard University Press, 1958.

Nicholson, William. "A remarkable effect of the inflection of light passing through wire cloth, not yet clearly explained." *Nicholson's Journal* 1 (1979): 13–16.

–. "Experiments and remarks on certain ranges of colours hitherto unobserved, which are produced by the relative position of plane glasses with regard to each other." *Nicholson's Journal* 2 (1799): 312–315.

[–]. "Light." *The British Encyclopedia*. Ed. by W. Nicholson. 6 vols. London: Longman et al., 1809. See also v. 7 of the American edition. 12 vols. Philadelphia: Mitchell et al., 1818.

Parrot, Georg. "Drei optische Abhandlungen: Die Theorie der Beugung des Lichts; die Theorie der Farbenringe; und über die Geschwindigkeit des Lichts." *Annalen der Physik* 51 (1815): 245–321.

–. *Entretiens sur la physique*. 6 vols. Dorpat: J. Schumann, 1819–24.

Pav, Peter Anton. "Eighteenth-century optics: the age of unenlightenment." Ph. D. diss., Indiana University, 1964.

Peacock, George. *Life of Thomas Young, M. D., F. R. S., etc., and One of the Eight Foreign Associates of the National Institute of France* London: J. Murray, 1855.

Péclet, Eugene. *Cours de physique*. Marseiile: A. Ricard, 1823.

–. *Traité élémentaire de physique*. 2nd ed. Paris L. Hachette, 1832.

Pemberton, Henry. "A Letter . . . concerning the abovementioned appearance in the rainbow, with some other reflections on the same subject." *Phil. Trans.* 32 (1722): 245–61.

The Philosophical Magazine. Vols. 7 (1800), 2, 8 and 12 (1801 to 1802).

Poisson, Simon-Denis. "Mémoire sur la théorie du son." *Journal de l'École Polytechnique* 7, cahier 14 (1808): 319–92.

–. "Mémoire sur le mouvement des fluides élastiques dans les tuyaux cylindriques, et sur la théorie des instruments à vent." *Mém. Acad.* 2 (1817): 305–402 (published in 1819). See also *Annales de Chimie* (1818): 288–99.

–. "Extrait d'un mémoire sur la propagation du mouvement dans le fluides élastiques." *Annales de Chimie* 22 (1823): 246–270; and "Lettre à Fresnel." Ibid, 270–80.

–. "Note sur le phénomène des anneaux colorés." *Annales de Chimie* 22 (1823): 337–48.

–. *Théorie mathématique de la chaleur*. Paris: Bachelier, 1835.

Potter, Richard. "On the modification of the interference of two pencils of homogeneous light." *The Philosophical Magazine*. 2 (1833): 81–94.

–. "A reply to the remarks of Professors Airy and Hamilton on the paper upon the interference of light." *The Philosophical Magazine* 2 (1833): 276–281.

–. "Particulars of a series of experiments and calculations undertaken with a view to determine the velocity with which light traverses transparent media." *The Philosophical Magazine* 3 (1833): 333–42.

–. "Mathematical considerations on the problem of the rainbow, shewing it to belong to physical optics [1835]." *Trans. Cambr. Phil. Soc.* 6 (1838): 141–52.

Pouillet, Claude. *Leçons de physique de la Faculté des Sciences de Paris.* 2 vols. Paris: Grosslin, 1828.

–. *Élémens de physique expérimentale et de météorologie.* 2 vols. Paris: Bechet, 1827–1830.

Powell, Baden. *Short Elementary Treatise on Experimental and Mathematical Optics.* Oxford: D. Talboys, 1833.

–. "Further remarks on experiments relative to the interference of light." *The Philosophical Magazine* 1 (1832): 433–38.

–. "Remarks on Mr. Barton's reply respecting the inflexion of light." *The Philosophical Magazine* 2 (1833): 424–34; and "Remarks on Mr. Barton's 'Reply . . .'" Ibid., 412–16.

Prevost, Pierre. "Quelques remarques d'optique, principalement rélatives à la réflexibilite des rayons de la lumière." *Phil. Trans. 1798: 311–31.*

Prieur, Claude. "Considerations sommaires sur les couleurs irisées des corps réduits en pellicules minces." *Annales de Chimie* 61 (1807): 154–79.

Priestley, Joseph. *The History and Present State of Discoveries Relating to Vision, Light, and Colours.* London: J. Johnson, 1772. Reprint. Millwood, N. Y.: Kraus Reprint Co., 1978.

Rittenhouse, David. "An optical problem, proposed by Mr. Hopkinson, and solved by Mr. Rittenhouse." *Transactions of American Philosophical Society* 2 (1786): 201–6.

Robison, John. *A System of Mechanical Philosophy.* 4 vols. Edinburgh: J. Murray, 1822.

–. "Acoustics;" "Sound;" "Trumpet;" and "Optics." *Encyclopaedia Britannica.* 18 vols. 3rd ed. Edinburgh, 1797.

–. "Temperament of the Scale of Music;" and "Musical Trumpet." Supplement to the *Encyclopaedia Britannica* 3rd ed. Edinburgh, 1801.

Ronchi, Vasco. *The Nature of Light: An Historical Survey.* Rev. and transl. from the Italian by V. Barocas. London: Heinemann, 1970.

Rossi, B. *Optics.* Reading, Mass.: Addison-Wesley Publ. Co., 1957.

Sauveur, Joseph. "Sur la determination du son fixe." *Hist. Acad. Paris.* 1700: H 161–78.

–. "Système générale des intervalles des sons, & son application à tous les systèmes et à tout les instruments de musique." *Hist. Acad. Paris* 1701: M 390–482.

Schweber, Sylvan S., ed. *Aspects of the Life and Thought of Sir John Frederic Herschel.* Vol. 1. New York: Arno Press, 1981.

Shapiro, Alan E. "Kinematic optics: A study of the wave theory of light in the seventeenth century." *Archive for History of Exact Sciences* 11 (1973): 134–266.

Silliman, Robert Horace. "Augustin Fresnel (1788–1827) and the establishment of the wave theory of light." Ph. D. diss., Princeton University, 1968.

Smith, Robert. *A Complet System of Optics in Four Books, vis. a Popular, a Mathematical, a Mechanical, a Philosophical Treatise.* Cambridge, 1738.

–. *Harmonics or the Philosophy of Musical Sounds.* 2nd ed., impr. and augm. London: Merrill, 1759.

Steffens, Henry John. *The Development of Newtonian Optics in England.* New York: Science History Publications/USA, 1977.

Stuewer, Roger H. "A critical analysis of Newton's work on diffraction." *Isis* 61 (1970): 188–205.

Stokes, Sir George Gabriel. *Mathematical and Physical Papers.* 5 vols. Cambridge: Cambridge University Press, 1880–1905. Reprint. New York: Johnson Reprint Corp., 1966.

Tenquist, D. W., R. M. Whittle, and J. Yarwood. *University Optics.* 2 vols. London: Iliffe Books, 1969.

Truesdell, Clifford. *The Rational Mechanics of Flexible or Elastic Bodies, 1638–1788.* Vol. 11 in 2nd ser. of *Leonhardi Euleri Opera Omnia.* Turin, 1960.

Tunbridge, Paul A. "An unpublished paper on light by Lord Henry Brougham F. R. S. (1778–1868)." *Archives des Sciences* 26 (1973): 111–17.

Venturi, Giovanni Battista. *Commentari sopra la storia e le theorie dell' ottica.* Vol. 1. Bologna: Fratelli et al., 1814.

Verdet, Émile. *Œuvres de E. Verdet.* 8 vols. Paris Imprimerie nationale, 1868–1872.

–. "Sur l'explication du phénomène des couronnes." *Annales de Chimie* 34 (1852): 129–40.

–. "Étude sur la constitution de la lumière non polarisée et de la lumière partiellement polarisée. *Annales scientifiques de l'École Normale Superieure* 2 (1865): 291–316.

Vieth, Gerhard. "Über Combinationstöne. In Beziehung auf einige Streitschriften über sie zweier englischer Physiker, Th. Young und Jo. Gough." *Annalen der Physik* 21 (1805): 265–314.

Weber, Ernst Heinrich, and Wilhelm Weber. *Wellenlehre auf Experimente gegründet oder über die Wellen tropfbarer Flüssigkeiten mit Anwendung auf die Schall- und Lichtwellen.* Leipzig: G. Fleischer, 1825.

Weber, Wilhelm. "Über Unterbrechungen der Schallstrahlen." *Journal für Chemie und Physik* 48 (1826): 385–430.

The Wellesley Index to Victorian Periodicals 1824–1900. Ed. by Walter E. Houghton. Toronto: University of Toronto Press, 1966.

Whewell, William. *History of the Inductive Sciences, from the Earliest to the Present Time.* 2 vols. 3rd ed. with add. New York: D. Appleton, 1901. First publication. London: J. Parker, 1837.

Whittaker, Edmund. *A History of Theories of Aether and Electricity.* 2 vols. Rev. and enl. ed. London: T. Nelson, 1951–53. Reprint. New York: Humanities Press, 1973.

Wilson, David Ball. "The reception of the wave theory of light by Cambridge physicists (1820–1850): A case study in the nineteenth-century mechanical philosophy." Ph. D. diss., John Hopkins University, 1968.

Winter, Richard. "A method of finding the specific gravity of light from analogy: and the undulatory system defended by an experiment in inflected light." *Nicholson's Journal* 19 (1808): 143–46.

Wollaston, William Hide. "A method of examining the refractive and dispersive powers, by prismatic reflections." *Phil. Trans.* 1802: 365–80.

–. "On the oblique refraction of Iceland crystal." *Phil. Trans.* 1802: 381–86.

–. "Description of a single-lens micrometer." *Phil. Trans.* 1813: 119–22.

Wood, Alexander. *Thomas Young Natural Philosopher, 1773–1829.* Completed by Frank Oldham. Cambridge: Cambridge Univ. Press, 1954.

Worrall, John. "Thomas Young and the 'refutation' of Newtonian Optics: a case-study in the interaction of philosophy of science and history of science." In C. Howson, ed. *Method and Appraisal in Physical Sciences.* Cambridge: Cambridge Univ. Press 1976.

–. "Fresnel, Poisson and the white spot: the role of successful predictions in the acceptance of scientific theories." In D. Gooding et al. (eds), *The Uses of Experiment.* Cambridge: Cambridge University Press, 1989.

Young, Matthew. *An Enquiry into the Principal Phenomena of Sounds and Musical Strings.* London: G. Robinson, 1784.

Young, Thomas. "Observations on vision." *Phil. Trans.* 1793:

–. "Outlines of experiments and inquiries respecting sound and light." *Phil. Trans.* 1800: 106–50.

–. "Esquisse d'une suite d'expériences et des recherches sur le son et la lumière; par Th. Young . . . *Bibliotheque Britannique* 14 (1800): 301–29.

—. "Lettre du Dr. Thomas Young . . . sur les découvertes faites par Aristote sur le son; et sur une formule pour les réfractions." *Bibliothèque Britannique* 18 (1801): 354–63.

—. "A letter from Thomas Young . . . respecting sound and light, and in reply to some observations of Professor Robison . . ." *Nicholson's Journal* 5 (August 1801): 161–67.

—. "On the mechanism of the eye." *Phil. Trans.* 1801: 23–88.

—. " 'Sur le mechanisme de l'œil par Thomas Young." *Bibliothèque Britannique* 18 (1801): 225–58.

—. *A Syllabus of a Course of Lectures on Natural and Experimental Philosophy.* London: the press of the Royal Institution, 1802.

—. "The Bakerian lecture. On the theory of light and colours." *Phil. Trans.* 1802: 12–48.

—. "An account of some cases of the production of colours, not hitherto described." *Phil. Trans.* 1802: 387–97.

—. "An answer to Mr. Gough's essay on the theory of compound sounds." *Nicholson's Journal* 2 (1802): 264–67.

—. "In reply to Mr. Gough's letter . . . on the phenomena of sound." *Nicholson's Journal* 3 (1802): 145–46.

—. "Observations in reply to Mr. Gough's letter on the grave harmonics." *Nicholson's Journal* 4 (1803): 72–73.

—. "Description of an apparatus for exhibiting the colours of thin plates by means of the solar microscope." *Journals of the Royal Institution* 1 (1802): 241–45.

—. "An account of Dr. Young's harmonic sliders." *Journals of the Royal Institution* 1 (1802): 261–64.

—. "The Bakerian lecture. Experiments and calculations relative to physical optics." *Phil. Trans.* 1804: 1–16.

—. "Untersuchungen über Schall und Licht, von Thomas Young . . . Bearbeitet von Director Vieth." *Annalen der Physik* 22 (1806): 249–85, 337–96.

—. *A Course of Lectures on Natural Philosophy and the Mechanical Arts.* 2 vols. London: J. Johnson, 1807. Reprint. New York: Johnson Reprint Corp., 1971.

—. "Hydraulic investigations, subservient to an intended Croonian Lecture on the motion of the blood." *Nicholson's Journal* 22 (1809): 104–24.

—. "Über die Theorie des Lichts, nach dem Systeme der wellenförmigen Schwingungen." *Annalen der Physik* 39 (1811): 156–205; "Nachricht von einigen Fällen einer bisher noch nicht beschriebenen Entstehung der Farben," ibid, 206–220; "Beschreibung einer Vorrichtung, um mittelst des Sonnen-Microscops die Farben dünner Flächen darzustellen," ibid, 255–61;

and "Versuche und Berechnungen zur physikalischen Optik," ibid, 262–84.

–. "On changeable colours and glories." *The Philosophical Magazine* 42 (1813): 292–96.

–. "Remarks on the measurements of minute particles, especially those of the blood and of pus." *The Annals of Philosophy* 2 (1813): 115–18, 190–95.

–. "Chromatics [1817]." In *Encyclopaedia Britannica*. Suppl. to 4th ed. Edinburgh, 1824.

–. "Theory of colours observed in the experiments of Fraunhofer." *The Edinburgh Journal of Science* 1 (1829): 112–16.

–. *Miscellaneous Works of the Late Thomas Young*. 3 vols. Editors: G. peacock (vols. 1–2), and J. Leitch (vol. 3). London: J. Murray, 1855. Reprint. New York: Johnson Reprint Corp., 1972.

–. The Royal Institution Notebooks. Library of University College, London, MS Add 13/14.

–. *A Reply to the Animadversions of the Edinburgh Reviewers, on Some Papers Published in the Philosophical Transactions*. London: Longman et al. & Cadell & Davis, 1804.

[–]. "Herschel on concentric rings between two object-glasses." *Retrospect of Philosophical . . . Discoveries* 4 (1809):

[–]. "Review of the *Mémoires d'Arcueil*." *The Quarterly Review* 3 (1810): 462–80.

[–]. "Review of Cuthbert's *New Theory of the Tides*." *The Quarterly Review* 6 (1811): 74–87.

[–]. "Review of Davy's *Elements of Chemical Philosophy*." *The Quarterly Review* 8 (1812): 65–84.

[–]. "Review of Malus, Biot, Seebeck, and Brewster on light." *The Quarterly Review* 11 (1814): 42–56.

–. Young correspondence. Royal Society, London, MS 242.

Index